Technological Transformation of GILDED AGE AMERICA

ANTHONY N. STRANGES

Kendall Hunt
publishing company

REVISED PRINTING

Cover image © Shutterstock, Inc.

Kendall Hunt
publishing company

www.kendallhunt.com
Send all inquiries to:
4050 Westmark Drive
Dubuque, IA 52004-1840

Copyright © 2014 by Anthony N. Stranges

ISBN 978-1-4652-6782-5

Kendall Hunt Publishing Company has the exclusive rights to reproduce this work,
to prepare derivative works from this work, to publicly distribute this work,
to publicly perform this work and to publicly display this work.

All rights reserved. No part of this publication may be reproduced,
stored in a retrieval system, or transmitted, in any form or by any
means, electronic, mechanical, photocopying, recording, or otherwise,
without the prior written permission of the copyright owner.

Printed in the United States of America

DEDICATION

For my grandchildren Victoria, Miranda, Giovanna, Madeline, and Dominic.

CONTENTS

PREFACE ix

CHAPTER 1: INTRODUCTION 1

CHAPTER 2: THE TECHNOLOGICAL TRANSFORMATION 5

    I. EUROPE AND ASIA 1870–1914 5

    II. THE RAPIDITY OF THE AMERICAN TECHNOLOGICAL TRANSFORMATION 6

      Factors that contributed to the American technological transformation in the last half of the nineteenth century 6

      The professionalization of science continued 7

    III. THE AMERICAN TECHNOLOGICAL TRANSFORMATION 7

      Steel 7

      Major contributors to the development of the large-scale steel industry, Kelly, Bessemer, and Mushet 8
      Significance of Kelly, Bessemer, and Holley 15
      An improved steelmaking process: the regenerative open-hearth furnace 15
      Major furnace advances 16
      United States steel production and cost and effect on the environment 18

Aluminum 24

Hall and Héroult, a case of nearly simultaneous invention 24

Petroleum 30

Origin of petroleum, early use as a medicinal and as an illuminant 32
First large petroleum discovery 35
Nineteenth-century methods of finding and drilling for petroleum 37
Petroleum storage and refining 40
Petroleum discoveries after Titusville 44
Thermal and high-pressure catalytic cracking of crude petroleum 45
United States petroleum reserves 46

Communications industry 50

Telephone 50

Electrical industry 61

Edison's contribution to the telephone 62
Light bulb 63
Establishing the General Electric Company 66
Westinghouse and the introduction of alternating current 68

Edison's other inventions 75

Phonograph 75
Motion picture machine 75
Spirit communication machine 75

A brief look at the emerging American chemical industry 82

Salt industry 82
Borax industry 83
Sulfur industry 84
Electrolytic industry: Dow Chemical Company 85
Plastics industry 85

Automobile industry 86

*External combustion engine: steam-powered and internal combustion
gasoline and diesel-powered vehicles* 86
American automobile manufacturers and organizers 88

**CHAPTER 3: INDUSTRIAL ORGANIZING METHODS OR TECHNIQUES
AND INDUSTRIAL ORGANIZERS** 91

I. **GENERAL CHARACTERISTICS OF MAJOR INDUSTRIES** 91

II. **LEADING INDUSTRIAL ORGANIZERS**............................92

III. **ORIGINS OF INDUSTRIALISTS AND ENTREPRENEURS
WHO DOMINATED AMERICAN INDUSTRY** 97

IV. **REACTION TO THE GROWTH AND CONSOLIDATION
OF INDUSTRY**..98

Public reaction..98

*Political reaction from state and federal governments: regulation
of railroads and corporations*....................................100

Regulation of railroads..100
Regulation of corporations.....................................101

**CHAPTER 4: ENVIRONMENTAL POLLUTION: THE OTHER SIDE OF
URBANIZATION, MODERNIZATION, AND
INDUSTRIAL DEVELOPMENT**............................103

I. **INTRODUCTION**...103

II. **WATER PURIFICATION AND THE PROBLEM OF WASTE
DISPOSAL IN THE UNITED STATES: THE THREE STAGES
OF DEVELOPMENT**..105

Background 105

Water distribution and sewer systems 110

Slow and rapid sand filtration systems 118

Chlorination 124

III.   **INDUSTRIAL WATER POLLUTION** 130

   IV.   **AIR POLLUTION** 131

           Acid rain and smoke. 132

           Cottrell and the electrostatic precipitator: the first industrial environmentalist. 138

CHAPTER 5: CONCLUSION. 141

       REVIEW QUESTIONS. 143

BIBLIOGRAPHY. 145

INDEX 151

PREFACE

Historians of science long have recognized that US history survey textbooks present a comprehensive account of the country's social, political, cultural, economic, and intellectual history but have not provided the history of its science and technology equal coverage. Social, political, cultural, economic, and intellectual developments are obviously important components of American history but so is its scientific and technological development. To claim that science and technology represent one of the most important, if not the most important, force that has contributed to the United States' global leadership is not an exaggeration. Jonathan Cole, a sociologist of science, Provost and Dean of Faculties at Columbia University, highlighted US history survey textbooks' lack of history of science and technology in a 1996 article he published in the *Bridge*, the US National Academy of Engineering's quarterly journal. Nothing has changed since that time except for the 2003 publication of Pauline Maier's *Inventing America*, a US history survey text that had two historians of science and technology among its four coauthors and included the advances that occurred in American science and technology. A second edition appeared in 2006 but has not undergone revision since that time.

Technological Transformation of Gilded Age America is not a comprehensive survey of American science and technology and has a limited objective. It examines the technological developments that occurred in the last half of the nineteenth century and thereby provides the missing coverage of Gilded Age technological developments so characteristic of US history survey textbooks. Significant technological development occurred at this time because the advances in nineteenth-century science enabled science to *catch up* to technology and provide a scientific explanation for the technological development that had occurred.

In the first half of the nineteenth century technology led science.[1] Innovators invented new products and made them work but did not understand the scientific laws or principles

---

[1] I first heard Mel Kranzberg (1917–95) use the phrase *technology led science* about thirty years ago at a meeting of ICOHTEC (The International Committee for the History of Technology). Kranzberg was one of the founders of the Society for the History of Technology (SHOT), long-time editor of its journal *Technology and Culture*, and a founder of ICOHTEC.

underlying their innovations. This began to change in the second half of the century because scientific research uncovered the scientific laws and principles that reduced or eliminated the trial-and-error, or hit-and-miss approach, characteristic of earlier inventors. Charles Goodyear's vulcanization of rubber process that he patented in 1844 represents the classic case of technology leading science. Goodyear found that heating naturally-occurring rubber with sulfur at a high temperature changed the chemical properties of rubber and significantly improved its quality, but he did not know why the change occurred.

Important scientific breakthroughs that contributed to science-based technological advancements occurred in the study of chemical composition, analysis, and reactions. They vastly improved the steel making process and turned it from an art to a science. Michael Faraday's laws of electrolysis and his discovery of electromagnetic induction, the relation between electrical current flow (ac, dc) in a conductor and heat generation, made possible the electrical and communication industries. The law of conservation of energy provided the scientific explanation for the energy changes that occurred in processes or inventions such as the conversion of sound energy to electrical energy, mechanical energy to heat energy, and electrical energy to heat energy. The recognition and application of scientific principles that advanced technological innovation is a major theme of this study.

Industrial organizers and the organizational techniques they employed and the environmental consequences of the Gilded Age technological transformation are other themes discussed in this study. Numerous scholarly works exist on the organizers and organization, and US history survey textbooks provide sufficient coverage of organizers such as Andrew Carnegie and John D. Rockefeller. They played an important role in the technological transformation that occurred, and their inclusion is essential to provide a complete account of the transformation that enabled the United States to emerge as an industrial giant. Until recently few publications examined the environmental consequences of the Gilded Age technological transformation, particularly water and air pollution. Most of the literature focused on the horrible working and living conditions that emerged, and justifiably so. A discussion of the environmental consequences therefore provides an additional look at the other not-so-pleasant side of the technological transformation. The realist writers of the Gilded Age, while not focusing on the environmental consequences, nevertheless provided the correct message for that time, namely, to describe nature, or the environment, as it really existed. *Technological Transformation of Gilded Age America* fulfills that requirement.

To recap, the three themes discussed are the new technological developments and their transforming impact on the United States, the organizers and the organizing techniques they employed and which made the United States the leading industrial nation in 1900, and the environmental consequences that resulted from the Gilded Age technological transformation.

The publication of this manuscript required considerable assistance from several people whom I wish to acknowledge at this time. Rita Walker, a staff member of the Texas A&M History Department, provided invaluable support in putting the manuscript in its final form. The Kendall-Hunt team, particularly Elizabeth Cray, Abby Davis, and William England, managed and directed the manuscript's editing and publication. I thank them for their help.

CHAPTER 1

INTRODUCTION

Chance favors only the prepared mind.
Louis Pasteur (1822–95)

The second half of the nineteenth century represented a transition from agricultural to industrial America. The change from steam power to electrical power drove the transformation. Technology led science in the first half of the nineteenth century, but after the Civil War the discovery and application of existing and new laws and principles in the physical sciences led to a technology that became an increasingly science-based discipline. The laws and principles included Faraday's laws of electrolysis, his principle of electromagnetic induction, James Joule's law relating heat and electrical current, and the laws of thermodynamics.

Nevertheless agriculture continued to flourish, and during the years of technological transition the number of farms and farmland acreage increased from 1,449,000 farms and 294 million acres in 1850 to 5,740,000 farms and 841 million acres in 1900. The total acreage, 850 million acres of commercial forests and 100 million acres of noncommercial forests, continued to fall, from 950 million acres in 1630 to 580 million acres in 1907. As industry, electrical generation, and irrigation grew so did water consumption grow and gradually outstrip population growth.

The number of patents awarded increased tremendously and represents the most striking example of the technological transformation occurring. The Patent Office issued 1,000 patents in the 1850s, 25,000 patents in 1890. Significant patents included those of Thomas Edison, Alexander Graham Bell, and Charles Martin Hall. Less significant patents included the 1856 foot seeder patent of George Meacham in New York City, the 1882 mustache guard patent of Charles Miller in Boonville, Missouri, and the 1884 patent awarded George Clark in Dubuque, Iowa, on a baby carriage that looked like a fancy high-top shoe.

Stock exchanges and stock indexes arose and expanded during the years of the technological transformation. The New York Stock Exchange, founded in 1792, received its present name in 1863. Its membership grew from 30 companies in 1893 to 200 companies in 1897 and included US Rubber and General Electric, both of which benefited because of the ongoing technological transformation. Despite the tremendous industrial growth and a trading volume that reached 1,000,000 shares per day in 1886, few Americans other than the wealthy actively participated in the stock market.

Charles Henry Dow (1851–1902), Edward D. Jones (1855–1920), and Charles Bergstresser (1858–1923), all of whom were financial reporters for New York City's Kiernan News Agency, in November 1882 founded Dow Jones & Company, a stock index and financial news service in New York City. On 8 July 1889 the three founders published the first edition of the *Wall Street Journal*. Dow served as editor. In 1884 Dow first compiled a stock average consisting of 11 stocks: Western Union, Pacific Mail Steamship, and nine railroad stocks. The Dow Jones Industrial Average made its debut on 12 May 1896 with a value of 40.94. It included 12 companies, among them General Electric, Chicago Gas, US Rubber, National Lead, American Sugar, and American Tobacco.

Throughout the decades of the technological transformation the United States experienced a series of significant recessions (contractions) and recoveries (expansions). From October 1873 to March 1879 it underwent a 65-month contraction, its longest up to that time, followed by a 36-month expansion from March 1879 to March 1882. Another 22-month expansion lasted from May 1885 to March 1887 followed by a shorter 13-month contraction from March 1887 to April 1888. Up to World War II expansions and contractions averaged about 30 months. Since World War II expansions have averaged about 53 months, contractions about 11 months.

BUSINESS CYCLE DATES			
(Roman numerals represent quarters, durations are in months)			
Peak	*Trough*	Contraction	*Expansion that Followed*
June 1857 (II)	December 1858 (IV)	18	30
October 1860 (III)	June 1861 (III)	8	22
April 1865 (I)	December 1867 (I)	32	46
June 1869 (II)	December 1870 (IV)	18	18
October 1873 (III)	March 1879 (I)	65	36
March 1882 (I)	May 1885 (II)	38	22
March 1887 (II)	April 1888 (I)	13	22
July 1890 (III)	May 1891 (II)	10	27
January 1893 (I)	June 1894 (II)	17	20
December 1895 (IV)	June 1897 (II)	18	18
June 1899 (III)	December 1900 (IV)	18	24
September 1902 (IV)	August 1904 (III)	23	21

From 1865 to 1886, during Reconstruction and the Gilded Age, many Americans made new fortunes in railroads, coal and mineral mining, and oil well production. The middle class prospered. The Third Great Awakening which followed and lasted from 1886 to 1901 experienced labor riots, agrarian protests, muckrakers, settlement houses, chautauquas, and evangelicals.

Laissez-faire and Social Darwinism dominated nineteenth-century economic and social-philosophical beliefs. Laissez-faire's proponents declared that government should not interfere with the workings of the economy, it should assist but not direct the economy. Social Darwinists held that those individuals or organizations able to adapt to their environment would survive, those unable to adapt would perish. Social Darwinists claimed that this was a law of nature, and therefore a law of God, and no one should interfere with its operation.

In the first half of the nineteenth century, technology led science. Inventors such as Samuel Morse (1791–1872) and Charles Goodyear (1800–60) usually did not understand the scientific principles and laws that underlay their inventions. Empirical thinking, spatial thinking, and serendipity guided or directed their work. The inventive process began to change in the second half of the century. Technology still led science, but science had advanced significantly and became increasingly essential to the inventive process. A recognition of the oxidation process in which oxygen from the atmosphere reacted with carbon present in pig iron led to the mass production of steel. An understanding of iron alloys and the acid-base reactions of furnace linings improved steel's quality. In the production of aluminum metal Charles Martin Hall's knowledge of electrolysis led to a commercial process for the separation of aluminum from its bauxite ore. In the petroleum industry understanding of reaction conditions, such as the use of high pressures and catalysts in refining, greatly improved the yields of products obtained from crude petroleum. Alexander Graham Bell's implicit recognition of the law of conservation of energy (sound → electrical → sound), of electromagnetic induction, and electrical resistance were an integral part of the telephone's invention. The same case holds for Edison's electrical inventions. Scientific laws and principles were transforming engineering into a science-based discipline that led to a technological transformation in late-nineteenth and early-twentieth-century United States.

This study examines the breakthroughs that contributed to the post-Civil War technological transformation and to the development of the steel, aluminum, petroleum, communications, and electric industries. The social and environmental impact of the technological transformation on American society and the organization of the steel, aluminum, petroleum, communications and electrical industries complete the examination of the transformation that occurred in the years 1850-1900.

CHAPTER 2

THE TECHNOLOGICAL TRANSFORMATION

I. EUROPE AND ASIA 1870–1914

Europe underwent a technological transformation at this time, although growth varied from country to country. England and Germany were the industrial leaders. England remained the longtime leader in the coal, steel, chemical, textile, and banking industries, but its plants and marketing techniques had become outmoded. Germany, having unified after the victorious Franco-Prussian War of 1870–71, emerged to challenge Britain, not only in the older established industries, but also in the lesser visible industries of banking, insurance, and shipping.

Other European countries, such as Russia, France, Italy, and Switzerland, advanced industrially. Russia, like the United States, possessed an abundance of raw materials, but its harsh weather conditions and lack of adequate land, water, and rail transportation facilities hindered industrial development. Russia had large industrial plants mixed with an abundance of small shops.

A major advance in Russian transportation occurred in 1891 when the Russian Czar Alexander III (1845–94) began construction of the Trans-Siberian Railroad, or *Iron Road*, the world's longest continuous railroad at 5,778 miles (9,198 km). It ran from the port city of Vladivostok on the Pacific Ocean to Moscow, or almost twice the distance of the United States railroad from New York City to San Francisco. When Alexander died in 1894 his son Nicholas II (1868–1918) continued the construction. Convicts and political prisoners made up part of the labor force that completed the construction in 1904. The *Iron Road* became the main entry way for new residents into Siberia. The trip required about eight days in transit.

France recovered relatively well from its defeat in the Franco-Prussian War, but its tariffs and protection policies retarded industrial growth. Switzerland and Italy developed high-quality, machine-manufactured goods, such as clocks and small-scale machines. Spain, Portugal, Greece, and Norway, despite its abundance of cheap water power, experienced little or no industrial development.

In Asia only Japan, following the Meiji restoration in 1868 and with assistance from western nations, such as Germany and Britain, began industrial-scale production of chemicals and electricity. China and the rest of Asia and Africa did not undergo industrial development at this time.

II. THE RAPIDITY OF THE AMERICAN TECHNOLOGICAL TRANSFORMATION

Factors that contributed to the American technological transformation in the last half of the nineteenth century

Money came from abroad. London, the world's financial capital, provided funds for investments in the United States. For the railroad industry between 1860 and 1880, probably half the capital, $1.4 billion in 1870 and $3.36 billion in 1900, came from overseas.

People, some of whom were inventors, came to the United States from abroad. Alexander Graham Bell came from Scotland to Canada and then to the United States in 1872. Michael Pupin came from Serbia in 1874, Nikola Tesla came from Croatia in 1884. The country's population grew from 38.5 million in 1870 to 50.1 million in 1880, 62.9 million in 1890, and to 75.9 million in 1900.

The Civil War had ended, construction replaced destruction. Natural resources, such as iron ore, coal, petroleum, and water, were plentiful. Transportation facilities, canals and railroads, in addition to the country's natural waterways, were widely available. The spirit of the times, a materialistic, entrepreneurial, risk-taking, confident outlook, prevailed. As a result of all these factors, from 1870 to 1910 production in the United States increased by about four percent annually. The United States' growth rate surpassed the growth rate in Germany, England, and France.

The professionalization of science continued

Scientists and engineers established several scientific and engineering associations and organizations from the 1850s to the 1920s. They sought to promote the nation's industrial growth through their professional organizations.

The chemical industry in 1862 established the Chemical Manufacturers Association (CMA) with its headquarters in Washington, DC. It was a nonprofit trade association of chemical manufacturers in the United States and Canada. The American Chemical Society (ACS), established in 1876 with its headquarters in Washington, DC, became one of the world's largest scientific societies. The American Institute of Chemical Engineers (AIChE), established in 1890 with its headquarters in New York City, promoted the advancement and development of chemical engineering.

Organizations that tested the quality and performance of materials arose. The American Society for Testing and Materials (ASTM), organized in 1898 and chartered in 1902 with headquarters in Philadelphia, established standards for the performance of materials. The American Society for Metals (ASM now ASM International) which existed under various names from 1913 until its formal establishment in 1935 with headquarters in Russell Township, Ohio, set standards for metal quality and performance.

Arthur D. Little (1863–1935) in Cambridge, Massachusetts, in 1886 founded one of the pioneering industrial research and chemical consulting organizations. His organization served as the prototype of American industrial consulting firms that have played a significant role in American industrial development.

III. THE AMERICAN TECHNOLOGICAL TRANSFORMATION

Steel

Steel, produced by heating naturally-occurring iron ores mainly hematite (Fe_2O_3) and charcoal obtained from the destructive distillation or carbonization of wood (trees), dates from 1000 BC. Egypt, Austria (ancient Noricum), and India in 500 BC were ancient centers of steel production. No mass production of steel occurred, and its manufacture remained an art rather than a science until the 1850s.

Abraham Darby (1678–1717), a Quaker industrialist in Coalbrookdale, England, in 1709 first used coke, obtained from the destructive distillation of coal, to replace the depleted supplies of increasingly expensive charcoal. In addition to its short supply charcoal presented a second problem because its softness prevented it from supporting a heavier column of iron ore in larger furnaces. Coke served as both a fuel and a reactant for the conversion of iron ore to wrought iron, steel, and pig or cast iron, but coke production from coal released nitrogen, sulfur, and mercury compounds present in coal and caused considerable air pollution. The presence of these compounds also made coal an unsuitable reactant in converting the ore to iron or steel because they introduced impurities and spoiled the product's quality. The use of coke, which copper and lead smelting industries already adopted, eliminated the introduction of these impurities.

Two other English inventors made important contributions to early steelmaking. Benjamin Huntsman (1704–66), a Sheffield clock maker, in 1745–50 invented the crucible steel process for making high quality steel, and in 1783 Henry Cort (1740–1800) in Fontley invented a grooved rolling mill to replace the laborious hammering of wrought iron or steel into sheets or plates.

What made steel different from wrought and pig iron became clear in 1786 when Claude Louis Berthollet (1748–1822), Gaspard Monge (1746–1818), and Alexandre Vandermonde (1735–96) in Paris showed the difference depended on the carbon content of each. Wrought iron contains 1–2 percent carbon, steel contains 2–4 percent carbon, and pig or cast iron contains more that 4 percent carbon.

In the United States an iron industry and small-scale steel manufacturing had been developing since colonial times. With the discovery of new iron ore deposits, such as the Mesabi Range in Minnesota in the early 1890s, iron and steel manufacturing in the Great Lakes region and in Pennsylvania continued to expand. In the south, Tredegar Iron Works, established in 1837 in Richmond, Virginia, was a major Confederate manufacturer during the Civil War. Its rolling mills produced heavy iron plates for the Confederate navy's ships including the *Merrimack*.

Major contributors to the development of the large-scale steel industry, Kelly, Bessemer, and Mushet

William Kelly (1811–88) was the son of Irish immigrants who became wealthy Pittsburgh landowners. He attended the Western University of Pennsylvania, renamed the University of Pittsburgh in 1908, where he studied metallurgy. Kelly worked in the steamboat industry until 1845 when a fire destroyed his property. In 1846 he moved to Eddyville in western Kentucky, where he and his brother John bought 14,000 acres of forested land containing iron ore, a blast furnace, and a Cobb charcoal wrought iron furnace. In Eddyville they built the Suwanee Furnace (Iron Works) and Union Forge to manufacture large black wrought iron kettles. Iron ore reacting at high temperature with coke in the coke-fired Suwanee blast furnace converted the ore to pig (cast) iron. At Union Forge, located a few miles from the Suwanee blast furnace, a two-stage furnace process in which Kelly heated pig iron in an air stream (the oxygen in air) converted pig iron to wrought iron. In the first, or refinery, stage a refinery hearth (called a run-out fire) using coke as the fuel to save charcoal, an air blast reduced the pig iron's carbon content by oxidizing some of the carbon and oxidized (removed) any impurities present such as sulfur, silicon, and manganese. In the second, or finery, stage a charcoal-fueled finery hearth (forge), also called a knobbing fire, oxidized any remaining pig iron and impurities in the wrought iron. The brothers sold the wrought iron kettles to plantation owners in Louisiana and Cuba who boiled sugar cane in them.

Kelly worried about his decreasing charcoal fuel supply that contributed to wrought iron's high production cost, about $82 per ton to convert pig iron to wrought iron. In 1847, while experimenting with his finery charcoal furnace, he noticed that air currents reacting with the molten pig iron in his furnace caused the pig iron to glow with a white heat. This chance observation led to two important conclusions. Kelly recognized that the oxygen from the air upon combining with the carbon in the pig iron to produce the desired low carbon-containing wrought iron released a tremendous amount of heat (an exothermic reaction) and that the heat produced by the reaction practically eliminated any external heating, resulting in a considerable reduction in fuel costs. The elimination of any external heating was a major claim of the iron manufacturing patent he received in 1857.

Four years of experimenting followed, and in 1851 Kelly constructed his first air-blown, or pneumatic, furnace in which he carried out the refinery and finery stages in a single operation. It was a square brick structure four feet high with a cylindrical chamber whose bottom had three holes (tuyeres) through which he blew air into the chamber containing molten pig iron. His innovative furnace design with its upward directed air blast prevented the molten iron from running into the holes and clogging them. A second furnace of the same design was 5 feet high with a 1.5 foot diameter. Kelly believed his furnace design was the most effective way of reacting air with pig iron. It was the first of several breakthroughs that led to the emergence of an international steel industry. A later breakthrough, the addition of limestone (calcium carbonate, $CaCO_3$) when heating the pig iron in a blast furnace, removed silicate impurities as slag and improved the steel's quality.

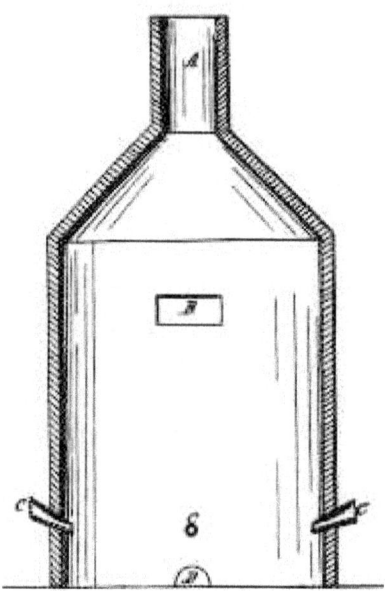

Figure 1   Kelly's Pneumatic Furnace

Kelly built seven pneumatic furnaces (converters) between 1851 and 1856, hoping to improve his Suwanee wrought iron making process. In 1854 he had eight charcoal pneumatic furnaces and one steam-powered hammer capable of producing about 1,000 tons of wrought iron blooms (lumps) per year. Until 1854 he rented about 300 slaves from their owners to work at Suwanee but this proved unproductive, so he imported 10 Chinese workers as replacements. Their importation made Kelly one of the first American industrialists to use Chinese workers.

Although Kelly invented his air blowing process in 1851, he did not patent it until 23 June 1857 while still in Kentucky (*Improvements in the Manufacture of Iron*, US Patent 17,628) and only after reading in a newspaper account that Henry Bessemer in England already had received a United States patent on steel making in 1856. Fortunately for Kelly the United States Patent Office, in a highly controversial decision, acknowledged his prior use of the air blast in furnace design.

The Suwanee Iron Works and the Panic (Depression) of 1857 left Kelly bankrupt, and in 1857 he left Kentucky for Pennsylvania. Kelly intended to improve his pneumatic furnace process, and from 1857–58 and in 1862 he conducted furnace tests at the Cambria Iron Works in Johnstown, Pennsylvania, a region with abundant coal and iron ore deposits. His cast iron furnace had a firebrick lining and an iron capacity of 1,500 pounds. The Cambria Iron Works, chartered in 1852 in Johnstown, then a village of 1,300 inhabitants, emerged from the consolidation of four primitive charcoal furnaces that were operating around Johnstown as early as 1809. The iron works experienced rough times because of the 1857 Panic and an 1857 fire that destroyed a large furnace. Largely through the efforts of Daniel Morrell (1821–85), who became Cambria's general manager in 1855 and later served in the US Congress, the iron works survived and beginning in the 1870s became a successful steel producer.

Kelly's pneumatic furnace process, as the title of his patent indicated, produced wrought iron, not steel. The pig iron-air reaction was too fast to control and removed too much carbon from pig iron, giving wrought iron instead of steel. Adding charcoal in a separate step (cementation process) converted the wrought iron to steel. But no efficient, large-scale process for converting wrought iron to steel existed until Robert Mushet (1811–91), an English metallurgist in Coleford and son of David Mushet (1772–1847) a famous Scottish iron and steel metallurgist, invented the spiegeleisen process in 1856. To improve the conversion of wrought iron to steel and give a better quality steel, Mushet, in 1848 began experimenting with spiegeleisen, his triple compound, an iron alloy containing 70–80 percent iron, 15–30 percent manganese, and a 4.5–6.5 percent carbon. Eight years later he patented the process of adding spiegeleisen to a pneumatic furnace to increase the carbon content of wrought iron and to remove excess oxygen gas that produced blowholes in steel, making steel brittle (*Improvements in the Manufacture of Iron and Steel*, British Patent 2,219; 22 September 1856 and US Patent 17,389; 26 May 1857). Mushet's process produced a high quality steel.

Despite Kelly's limited technological success at Cambria, financial success eluded him until 1863 when he joined a syndicate of ironmasters who named themselves the Kelly Pneumatic Process Company. The Detroit multimillionaire shipping and lumber magnate Eber Brock Ward (1811–75) and the Massachusetts cousins, manufacturer Zoheth Durfee (1831–80) and engineer William Durfee (1832–99), bought Kelly's patent and organized the syndicate to develop Kelly's process. Kelly held a 30 percent interest in the syndicate. The syndicate secured the rights to Mushet's 1857 United States patent on 24 October 1864 and as part of the agreement brought Mushet and his two partners (John Brown and Thomas Clare) into the syndicate. The syndicate constructed the Eureka Iron Works in Wyandotte, near Detroit, and began commercial production of steel using a 2.5 ton converter in September 1864. Kelly allegedly earned $30,000 in royalties from his first (1857) patent, and its renewal in 1871 earned $450,000 in royalties, but uncertainty remains about the amount and whether Kelly or the syndicate received the royalties. Kelly spent his last years in Louisville, Kentucky, manufacturing axes and working as a realtor and banker.

Henry Bessemer (1813–98), the son of an English engineer and the inventor of movable dies for embossed stamps, gold print, and sugar machinery, developed his steel-making process in Sheffield. Prior to his investigations, England in 1850 produced 2.5 million tons of iron but only 60,000 tons of steel. At the time of the Crimean War of 1854–56 in which Britain and France were fighting Russia, Bessemer tried to make a cannon structurally strong enough to fire a rifled projectile that would spin in flight and keep a more stable trajectory. Such a cannon would fire farther and more accurately. He failed to interest the British military or the French emperor Napoleon III (1808–73) who doubted the strength of Bessemer's cannon. Bessemer acknowledged the weakness of a cannon cast of iron or brass, a copper-zinc alloy, but believed in the strength of a steel cannon. Steel was very expensive, so he looked for a cheap production process.

Bessemer reacted iron ore with coke to give pig iron. Like Kelly, Bessemer knew that a blast of air blown up through molten pig iron in an iron converter lined with firebrick considerably reduced pig iron's carbon content giving wrought iron. His first significant contribution to the steelmaking process, the Bessemer rotatable converter mounted on trunnions, simplified the loading of reactants and unloading of products. His second significant contribution followed from his converter tests when he noticed that the heat produced from the reaction of carbon in the pig iron with oxygen from the air kept the temperature very high and eliminated any external heating. Bessemer's process, like Kelly's, produced wrought iron because the reaction was too fast to control and removed too much carbon from pig iron giving wrought iron instead of steel. Bessemer then added charcoal in a separate step (the cementation process) to convert the wrought iron to steel.

Bessemer, on 17 October 1855, received a British patent (British Patent 2,321) and on 11 November 1856 received a US patent (US Patent 16,082) entitled *Manufacture of Iron and Steel*. In 1860 he gained free use of Mushet's spiegeleisen process, enabling him to improve his steel making process immensely. Because Mushet could not finance the facilities and equipment to continue his experiments he assigned a 50 percent share of his patent to his associate Thomas Brown, managing partner of the Ebbw Vale Iron Company a large iron works in South Wales and London. A twenty-five percent share went to another associate S. H. Blackwell, and the remaining 25 percent to himself. Blackwell and Brown, who were the patent's trustees, neglected to pay the patent's third year patent stamp duty of £50 ($245) or to inform Mushet of their neglect. As a result Mushet's 14-year patent lapsed in 1859, making it public property and giving free use to British iron and steel industrialists, particularly Bessemer. Both Mushet and Bessemer failed to convince the US Patent Office to renew their fundamental patents for another seven years when they expired in 1870.

By 1890 Bessemer had received over one million pounds (about $5 million, 1£ = $5) in royalties on his steel production patents, calculated at the rate of 1£ per ton of steel. Following the lapse of his patent Mushet appealed to Bessemer for a royalty payment. Bessemer refused, although for more than 20 years he paid Mushet a meager £300 ($1,500) annual pension that totaled over £7,000 ($35,000) at the time of Mushet's death in 1891. Bessemer also supported Mushet's receipt of the Iron and Steel Institute's Bessemer Medal in 1876, a small gesture compared to the millions in monetary compensation Mushet never received. Mushet believed £500 annually was a fair value of the royalties he should have received. The only patent royalty Mushet earned came from his US patent from which he received the sixteenth part of 1/32 of a share, a few hundred pounds, before it ran out and before the steel industry expanded significantly in the United States.

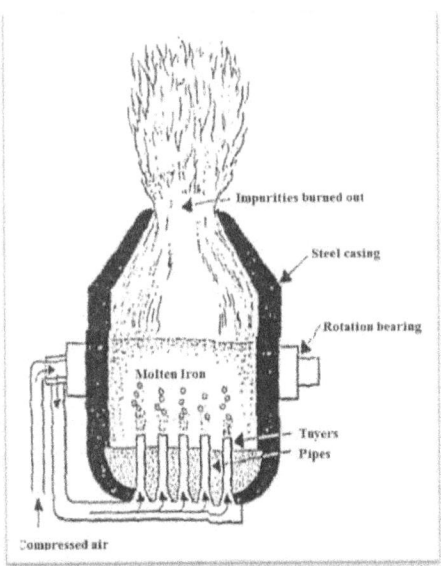

Figure 2   Bessemer Process for Steel Production

Commercial production of steel with Bessemer's converter began in 1859 in Sheffield, and as early as 1861 steel rails were in use in England. Production cost was low, £7 per ton (1£ = $5), compared to the earlier steel production cost of £22 per ton for low-quality, puddled steel and £100 per ton for high-quality, crucible steel, and comparable to the cost of wrought iron at £4 per ton. The Bessemer process spread to France in 1858, Germany in 1862, and Austria in 1863. By 1900 a large Bessemer converter produced 15–20 tons per batch at the rate of about one ton per minute.

Alexander Holley (1832–85), an engineer-metallurgist and technical writer in New York City, with the financial backing of two Troy, New York, businessmen, bought Bessemer's United States patent rights in 1863 for £10,000 ($70,800). In 1864 Holley built a second American Bessemer plant in Troy. The plant had a 2.5 ton converter and began operation in February 1865. That same year Holley completed a Kelly-Bessemer patent merger, naming the new organization The Trustees of the Pneumatic or Bessemer Process of Making Iron and Steel in Troy; renamed and reorganized as the Pneumatic Steel Association in New York City in 1866; then the Bessemer Steel Company or Association in Philadelphia in 1877; and finally the Steel Patents Company in Philadelphia 1890. The Bessemer group controlled 70 percent of the new company and the Kelly group the remaining 30 percent. Bessemer had patented the rotating converter in 1856, Kelly patented the air blast that he used in the process prior to Bessemer in 1857, and each needed the other's invention for the process to succeed.

Holley built another Bessemer plant in Harrisburg, Pennsylvania, with two seven-ton converters that began operation in June 1867. In August 1867 the Cambria Iron Works, the sixth Bessemer plant in the United States, produced the first Bessemer steel rails on a commercial-scale and sold them for $104 per ton, considerably less than $170 per ton cost of imported English steel rails. It had two six-ton converters, each increased to 11.5 tons. Located in Pennsylvania's abundant coal and iron ore region and with easy access to water and rail transportation, the Cambria Iron Works in 1876 became the largest rail producer in the United States, producing 10 percent of the nation's rails and employing 18,000 men at its peak.

On 1 June 1889 Johnstown, now a growing steel company town of 30,000 inhabitants mainly German and Welsh, experienced the worst flood in United States history. Established on a flood plain in 1794 at the fork of two rivers, Johnstown's poorly built and maintained nearby South Fork Dam gave way, releasing 20 million tons of water that roared downhill at 40 miles per hour wiping out the city in ten minutes. Over 2,200 died. Full recovery from the flood took five years.

Significance of Kelly, Bessemer, and Holley

Both Kelly's and Bessemer's inventions were in part accidental, but they were observant enough to make two important conclusions. First, the oxygen in the atmosphere converted pig iron to wrought iron by reacting with (oxidizing) and removing the carbon in pig iron as carbon dioxide gas. Second, the reaction produced enough heat to sustain itself and thereby eliminate the need for an external source of heat. Adding carbon as the spiegeleisen alloy converted the wrought iron to steel. Their process replaced the two older methods of either mixing the proper proportions of iron ore with pig iron to get steel, or of adding carbon to wrought iron to give steel. Neither of these processes used the air blast to convert large quantities of pig iron to wrought iron quickly and then convert the wrought iron to steel. They remained small-scale processes and did not lead to a large-scale steel industry.

KELLY-BESSEMER PRODUCTION OF STEEL

iron ore (iron + oxygen) + coke (C) → pig iron (iron + carbon) + carbon dioxide (CO_2)

pig iron (iron + carbon) + oxygen (O_2) → wrought iron (iron + carbon) + carbon dioxide (CO_2)

wrought iron + spiegeleisen → steel

An improved steelmaking process: the regenerative open-hearth furnace

In 1856–61, the German-born brothers Friedrich Siemens (1826–1904) and Karl Wilhelm (Charles William, 1823–83) in Birmingham, England, and independently Pierre E. Martin (1824–1915) in 1864 in Sireuil, France, invented the regenerative open-hearth furnace process for manufacturing steel. The process was slower and provided better temperature control than the Kelly-Bessemer process because the heat came from an external source rather than internally. In the regenerative type open-hearth furnace, which provided higher temperatures and conserved fuel, the producer gas (a nitrogen, hydrogen, and carbon monoxide mixture) used as fuel and the air required for combustion were preheated before entering the furnace by passing them through a checkerwork of hot firebrick. After passing over and heating the reactants in the furnace, the hot waste gases left the furnace through a second firebrick, heating the firebrick. Furnace operators reversed valves that controlled the direction of the fuel gases and the waste gases every twenty minutes, keeping both checkerworks hot. The open-hearth furnace produced greater quantities of better quality steel, 50–200 tons per batch in 6 to 12 hours.

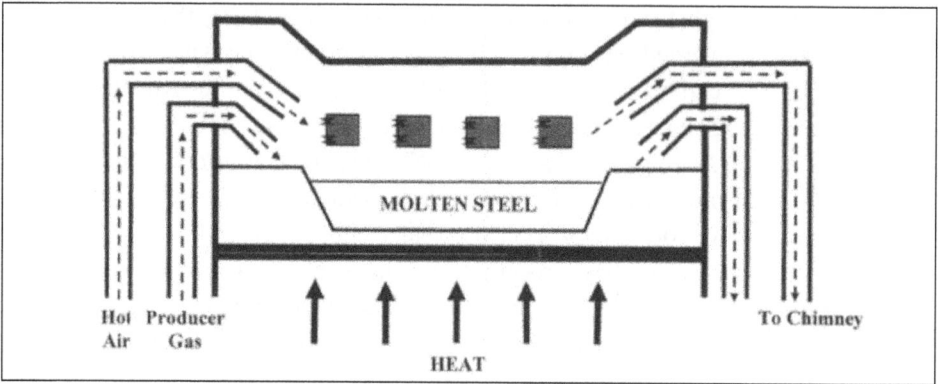

Figure 3 Open Hearth Process for Steel Production

Martin heated as his reactants, wrought iron and pig iron, to get the correct carbon content for steel. The Siemens heated iron ore and pig iron. Friedrich Siemens in 1861 developed a coal gasifier for manufacturing the inexpensive producer gas for fuel. Twentieth-century manufacturers use melted iron, scrap steel, iron ore, and limestone as reactants. Beginning in 1884 open hearth furnaces had either a dolomite lining (calcium magnesium carbonate, $CaCO_3 \cdot MgCO_3$) if the iron source had phosphorus (acidic) impurities, or a silica lining (silicon dioxide, SiO_2, present in quartz and sand) if the iron source had basic (manganese) impurities. The lining removed the impurities as a slag.

Cooper, Hewitt, and Company in 1868 built the first open-hearth furnace in the United States in Trenton, New Jersey. By 1900 open-hearth steel production exceeded Bessemer steel production, and by 1920 it replaced the Bessemer process as the main steel producer. The open-hearth process up to 1992 accounted for about 85 percent of total steel production in the United States and globally.

Major furnace advances

The first Bessemer and open-hearth furnaces had silica linings and were suitable only for iron ore that did not contain any phosphorus oxide (phosphorus pentoxide, P_4O_{10}) impurity. This was a serious problem for Bessemer because English iron ores usually contained phosphorus oxide. Sidney Gilchrist Thomas (1850–85) and his cousin Percy Gilchrist (1851–1935) in South Wales and later Middlesbrough,

England, solved the problem. Their use of a dolomite furnace lining, which they patented in 1877–78, eliminated phosphorus oxide present as an impurity in iron ore. The acidic phosphorus oxide reacted with the basic carbonate lining to form a slag that workers poured off the top of the molten steel. The slag when ground to a powder found use as phosphate fertilizer.

Iron ore containing phosphorus had plagued the process because phosphorus formed phosphate solids that remained mixed in the steel, resulting in a very low quality steel, whereas the steelmaking reaction removed any sulfur present in iron ore as sulfur dioxide gas. Thomas and Gilchrist's invention opened up the vast iron ore fields of Lorraine, France, and Luxemburg. Andrew Carnegie in 1880 paid Thomas and Gilchrist $300,000 for the United States rights to use their patents in his steel plants.

James H. Darby (1856–1919), the managing director of the Brymbo Iron and Steel Works near Wrexham, North Wales, first used a dolomite (basic) firebrick lining in the open-hearth furnace to remove phosphorus from iron ores in 1884. The cast iron open-hearth furnaces were of 12 and 20 tons capacity and produced 180–200 tons of soft steel ingots per week. Both Bessemer and open-hearth furnaces had dolomite or silica linings, depending on the type of impurity present in the iron sources used as the reactant. Specialty steel alloys first appeared in the late nineteenth century.

SPECIALTY STEEL ALLOYS		
Inventor	*Type of Steel*	*Properties and Uses*
Robert Hadfield (1858–1940), English metallurgist in Sheffield	Manganese (9–13 percent) in 1882	Made steel tough, used in armor plate for ships and tanks.
Charles E. Guillaume (1861–1938), Swiss physicist in Paris	Nickel (40 percent) in 1896	Addition of nickel gave a low-coefficient of expansion steel suitable for use in clock pendulums and surveyor's tapes. Invar is an example of this kind of steel. In 1920 Guillaume received the Nobel Prize in physics for his work on iron-nickel alloys.
Leslie Aitchison (1891–1973), English metallurgist in Ettington	Tungsten (0.5–3 percent) and molybdenum (0.24–8 percent) in 1900	Produced steel suitable for high-speed cutting tools.
Henry E. Brearley (1871–1948), metallurgist at the Brown Firth Laboratories in Sheffield	In 1913, while trying to produce a suitable alloy for gun barrels, he accidently invented a rust-resistant, chemical (acid)-resistant stainless steel containing 12.8 percent chromium.	Industries that manufacture armor and armor piercing projectiles use chromium steels because they are hard and tough. Food, drug, and chemical industries use steels containing 8–18 percent nickel and chromium.

United States steel production and cost, and effect on the environment

After the Civil War steel became the fourth largest industry behind lumber, flour, and textiles, helped in part by tariffs of 45 percent and as much as 100 percent on British steel. Steel replaced iron in rails by 1900 because it was seven times more durable, it also found wide use in skyscrapers and other buildings, bridges such as the Brooklyn Bridge which opened in 1883, and tools. From a production of only 20,000 tons of steel in 1866, production in 1876 increased to 597,000 tons and to 11.277 million tons in 1900, compared to a global production of 28.273 million tons. British steel production for this period was 292,000 tons in 1870, 1.375 million tons in 1880, 3.677 million tons in 1890, and 5.050 million tons in 1900. Before the introduction of the Kelly-Bessemer process, steel cost $200 per ton in 1810 and as much as $300 per ton. A few years after its introduction the selling price of steel dropped from $28 per ton in 1870 to $11.50 per ton in 1900.

STEEL PRODUCTION IN THE UNITED STATES			
Year	Tons (in thousands)	Year	Tons (in thousands)
1860	13	1885	1,917
1866	20	1890	4,779
1870	77	1895	6,785
1871	82	1900	11,277
1872	160	1905	21,880
1873	223	1910	28,330
1874	242	1915	35,180
1875	437	1920	46,183
1876	597	1925	49,705
1877	638	1930	44,591
1878	820	1935	38,184
1879	1,048	1940	66,983
1880	1,397		

The mass production of steel initiated the mass consumption of natural resources. One pound of steel required two pounds of iron ore, one to three pounds of coal for fuel and for conversion to coke, 0.33 pounds of limestone, and about 1.8 pounds of air for its production. The emergence of steel, and other late nineteenth-century industries, also represented the beginning of significant increases in air and water pollution resulting from the escape of carbon dioxide, nitrogen oxides, sulfur dioxide, and mercury into the atmosphere and the dumping of silicate and phosphate solid waste into rivers and lakes.

The amounts of carbon dioxide released into the atmosphere from fossil fuel burning had doubled every decade since colonial times, but the amounts of carbon dioxide released during the United States post-Civil War industrial expansion became increasingly problematic. The petroleum industry that began with the large-scale discovery at Titusville, Pennsylvania, in 1859, added to the atmospheric pollution.

A similar pattern developed in Europe's industrialized nations: England, Germany, France, and Italy; and in Japan. The earliest carbon dioxide measurements from England date from 1751, indicating the start of the eighteenth-century industrial revolution and introduction of the steam engine. Figures for the United States, Germany, France, Canada, Japan, China, and the now defunct Soviet Union's 15 republics provide useful comparisons.

	UNITED STATES			
Year	Total production of CO_2 (Thousand Metric Tons of Carbon)	CO_2 Production Gas Fuels	CO_2 Production Liquid Fuels	CO_2 Production Solid Fuels
1800	69	0	0	69
1810	114	0	0	114
1820	216	0	0	216
1830	570	0	0	570
1840	1,603	0	0	1,603
1850	5,402	0	0	5,402
1860	12,947	0	55	12,892
1870	26,916	0	559	26,357
1880	54,226	0	2,834	51,392
1890	109,647	3,190	4,855	101,603
1900	180,878	3,150	6,728	171,000

GERMANY

Year	Total production of CO_2 (Thousand Metric Tons of Carbon)	CO_2 Production Gas Fuels	CO_2 Production Liquid Fuels	CO_2 Production Solid Fuels
1792	128	0	0	128
1830	1,308	0	0	1,308
1865	17,322	0	10	17,312
1900	89,181	0	827	88,354

ENGLAND

Year	Total production of CO_2 (Thousand Metric Tons of Carbon)	CO_2 Production Gas Fuels	CO_2 Production Liquid Fuels	CO_2 Production Solid Fuels
1751	2,552	0	0	2,552
1800	7,269	0	0	7,269
1861	47,775	0	1	47,774
1900	114,558	0	834	113,724

FRANCE

Year	Total production of CO_2 (Thousand Metric Tons of Carbon)	CO_2 Production Gas Fuels	CO_2 Production Liquid Fuels	CO_2 Production Solid Fuels
1802	611	0	0	611
1830	1,788	0	0	1,788
1855	9,026	0	10	9,016
1900	35,283	0	201	35,082

ITALY				
Year	Total production of CO_2 (Thousand Metric Tons of Carbon)	CO_2 Production Gas Fuels	CO_2 Production Liquid Fuels	CO_2 Production Solid Fuels
1860	8	0	0	8
1864	438	0	1	437
1890	3,332	0	60	3,271
1902	4,098	1	60	4,037

JAPAN				
Year	Total production of CO_2 (Thousand Metric Tons of Carbon)	CO_2 Production Gas Fuels	CO_2 Production Liquid Fuels	CO_2 Production Solid Fuels
1868	3	0	0	3
1871	8	0	1	7
1890	1,845	0	116	1,729
1908	10,262	0	404	9,858
1916	16,602	8	418	15,636

CHINA				
Year	Total production of CO_2 (Thousand Metric Tons of Carbon)	CO_2 Production Gas Fuels	CO_2 Production Liquid Fuels	CO_2 Production Solid Fuels
1899	26	0	0	26
1911	7,600	0	0	7,600
1926	8,076	0	2	8,074

| EASTERN EUROPE |||||
Year	Total production of CO_2 (Thousand Metric Tons of Carbon)	CO_2 Production Gas Fuels	CO_2 Production Liquid Fuels	CO_2 Production Solid Fuels
1800	111	0	0	111
1858	2,340	0	1	2,339
1900	52,956	0	7,924	45,032
1919	25,246	70	5,155	20,021

| CENTRAL AND SOUTH AMERICA |||||
Year	Total production of CO_2 (Thousand Metric Tons of Carbon)	CO_2 Production Gas Fuels	CO_2 Production Liquid Fuels	CO_2 Production Solid Fuels
1884	1	0	1	0
1900	1,137	0	32	1,105
1915	4,902	3	922	3,976

| AFRICA |||||
Year	Total production of CO_2 (Thousand Metric Tons of Carbon)	CO_2 Production Gas Fuels	CO_2 Production Liquid Fuels	CO_2 Production Solid Fuels
1884	6	0	0	6
1901	1,067	0	0	1,067
1910	4,791	0	39	4,752

From 1751 to 1996 the burning of solid, liquid, and gaseous fossil fuels and cement production (the first figures are from 1927) released globally 265 billion metric tons of carbon dioxide into the atmosphere. In 1996 the combustion of solid and liquid fuels contributed 77.5 percent or 5,065 million metric tons of the global carbon dioxide emissions. Gas fuels, mainly natural gas, contributed 18.3 percent or 1,196 million metric tons. Gas flaring and emissions from cement production, which experienced a twenty-fold increase since the 1920s to 202 million metric tons, accounted for the remainder, 4.2 percent, of the total carbon dioxide emissions in 1996.

Aluminum

Hall and Héroult, a case of nearly simultaneous invention

Early attempts to separate aluminum metal from its naturally-occurring compounds using chemical and electrolytic processes were either impractical or expensive or both. The Danish physicist Hans Christian Oersted (1777–1851) in Copenhagen first produced aluminum in 1825 by reacting potassium amalgam (a potassium-mercury alloy) with anhydrous aluminum chloride and then distilling the mercury from the resulting aluminum amalgam, leaving an impure aluminum.

The German chemist Friedrich Wöhler (1800–82) at Göttingen first isolated pure aluminum in 1827 by chemically reducing (separating) the aluminum in white or yellowish crystalline aluminum chloride with potassium, a very reactive metal. Several years later in 1854, Robert Bunsen (1811–99), a chemist at Heidelberg, carried out the first electrolytic separation (reduction), electrolyzing molten sodium aluminum chloride (Na_3AlCl_6) to give aluminum. Bunsen used batteries because no large direct current (dc) generators were available for commercial-scale production, resulting in a high-production cost of about $140 per pound. In Paris the chemist Henri Sainte-Claire Deville (1818–81) in 1859 electrolyzed molten cryolite (Greenland Spar, Na_3AlF_6) and electroplated aluminum on copper.

Several years later, in 1884–85, Hamilton Castner (1858–99), an American chemist-chemical engineer, separated aluminum from aluminum chloride with sodium, another reactive metal. He obtained sodium relatively cheaply by reducing sodium in molten sodium hydroxide with iron carbide. American industrialists showed little interest in or willingness to provide financial backing for Castner's aluminum process, and in 1886 he moved from New York City to England where in 1888 he became managing director of the Aluminum Company established in 1887 in Oldbury. With the introduction of Hall's electrolytic process, Castner's chemical process could not compete economically. Instead he used the sodium, produced for

9¢ a pound, to manufacture sodium peroxide bleach and sodium cyanide, two compounds that the gold mining industry required.

In 1890 Castner invented an electrolytic process in which the electrolysis of brine (salt water) gave pure sodium hydroxide (caustic soda) and hydrogen and chlorine as by-products. Karl Kellner (1850–1905), an Austrian chemist and mystic-occultist, independently developed the same process, leading to a merger in 1895 of Castner's Aluminum Company and the Solvay Company in Brussels, Belgium, which held the patent rights to Kellner's process. The new company, called the Castner-Kellner Alkali Company, took advantage of inexpensive hydroelectric power generated by Niagara Falls and in 1896 constructed an alkali plant in Niagara Falls, New York. The following year, in 1897, the company constructed a large alkali plant in Cheshire, England.

The English alkali producer Brunner Mond absorbed Castner-Kellner Alkali in 1926 and merged with three other English chemical companies to form Imperial Chemical Industries (ICI) in London. Castner was a founding member of the Society of Chemical Industry established in 1881 which, since 1926, has awarded annually the Castner Medal for Electrochemical Research in his honor. Castner died quite young from tuberculosis.

No practical aluminum separation process existed when Charles Martin Hall (1863–1914), a minister's son and a third-year student at Oberlin College in Ohio, heard Frank Jewett (1844–1926), his chemistry professor, discuss the difficulty of separating pure aluminum from its ores and other compounds. Jewett had met Wöhler when in Germany in 1873–75, and after teaching at the Imperial University in Tokyo from 1876 to 1880, he returned to the United States to teach chemistry at Oberlin.

Hall already had an interest in chemistry and had read widely in chemistry textbooks and journals. He accepted the challenge of separating aluminum from its bauxite ore, an abundantly-available, naturally-occurring, reddish sedimentary rock that contained 30–75 percent alumina (aluminum oxide, Al_2O_3) and oxides of iron, silicon, and titanium. Chemists had succeeded in separating alumina from the other oxides by dissolving bauxite in hot concentrated sodium hydroxide to give aluminum hydroxide, cooling and precipitating the aluminum hydroxide, and then calcining (heating) it to give anhydrous (dry) alumina. They failed to develop a practical process for separating aluminum from its oxide.

Eight months after graduating in 1885, and after considerable experimenting in his backyard shed, Hall invented a direct current electrical process for the separation of aluminum on 23 February 1886. Unlike previous experimenters Hall dissolved alumina in cryolite (Na_3AlF_6) at 1,000°C, poured the molten solution into a small electrolytic cell that consisted of a graphite container and two graphite electrodes, and connected the electrodes to a 5–6 volt battery to start the electrolysis. Hall was searching for a way to lower the melting point of alumina, which was too high (2,030°C) and too impractical and uneconomical to maintain. Alumina also dissolved with great difficulty in most liquids including acids and strong alkalies. He found that dissolving alumina in cryolite cut the melting point in half and allowed him to use alumina rather than less abundant cryolite or aluminum chloride as his aluminum source.

Hall used graphite (carbon) because it did not interfere with the aluminum separation that he hoped to accomplish. A previous electrolysis using a clay container failed because clay, essentially a hydrated aluminum silicate sometimes containing iron oxide, interfered with the separation. The equations for the electrolysis below show the aluminum depositing on the cathode or negative electrode and carbon dioxide forming at the anode or positive electrode.

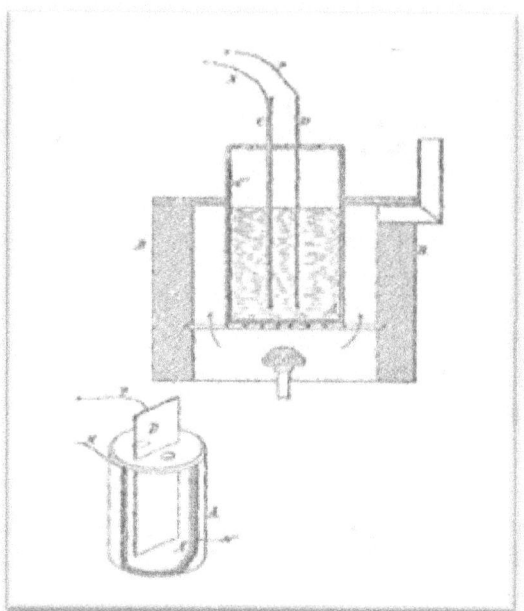

Figure 4   Hall's Patent on Aluminum Electrolytic Process

ELECROLYTIC PRODUCTION OF ALUMINUM	
Cathode (−)	$4\ Al^{3+} + 12\ e \rightarrow 4\ Al$
Anode (+)	$6\ O^{2-} - 12\ e \rightarrow 3\ O_2$ $3\ C + 3\ O_2 \rightarrow 3\ CO_2$

Hall's older sister Julia Brainerd Hall (1859–1926), who enrolled in most of the same science courses as her brother at Oberlin, worked with him. Her accurate laboratory records and the two descriptive letters dated 23 and 26 February 1886 that they sent to their brother George Hall in Dover, New Hampshire, helped Hall establish his priority of invention. Commercialization followed in November 1888 when Hall, with Alfred Hunt (1855–99) a Pittsburgh metallurgist and financial backer, and Arthur Vining Davis (1867–1962) a recent graduate of Amherst College, and $20,000 capitalization from the Mellon family, established the Pittsburgh Reduction Company in Pittsburgh, Pennsylvania. Hall, after three years of litigation, obtained a patent on his electrolytic production process on 2 April 1889 (filed 9 July 1886) and in 1890 became Pittsburgh Reduction's vice-president. Pittsburgh Reduction, which in 1890 increased its capitalization to $1,000,000, changed its name to Alcoa in 1907 and appointed Davis president in 1910.

Direct current dynamos (generators) that provided an electric current of several hundred amperes were available by 1886. This made commercial-scale aluminum production practical by connecting a large number of electrolytic cells in series and using a dynamo to generate the current required for the electrolysis. Hall constructed his first electrolytic cells, or pots, of cast iron, 24 inches long, 16 inches wide, and 20 inches deep (61 cm × 41 cm × 51 cm) and lined them with 3-inch (8 cm) baked graphite. Six to ten graphite anodes, 3 inches in diameter and 15 inches long (8 cm × 38 cm), hanging from a copper busbar above the cells, dipped into the molten alumina-cryolite solution. Each of the two electrolytic cells held 300–400 pounds (1 pound = 454 grams) of cryolite bath in which he dissolved alumina. A gas flame installed below the cells proved unnecessary because the heat produced by the two steam-powered dc dynamos connected in parallel kept the reactants and the aluminum product molten during electrolysis.

The large dynamos, each rated at 1,000 amperes and 25 volts, supplied 1,750 amperes at 16 volts to the two electrolytic cells connected in series that produced 50 pounds of aluminum per day. Enlarging the Pittsburgh plant increased production to 475 pounds per day in 1890. A new plant constructed in 1891 at New Kensington, 14 miles from Pittsburgh, produced 1,000 pounds per day in 1893 and

2,000 pounds per day in 1894. Pittsburgh Reduction constructed another plant in Niagara Falls, New York, in 1895. The General Electric Company, established in 1892 in Schenectady, New York, built the dynamos used in the larger aluminum plants.

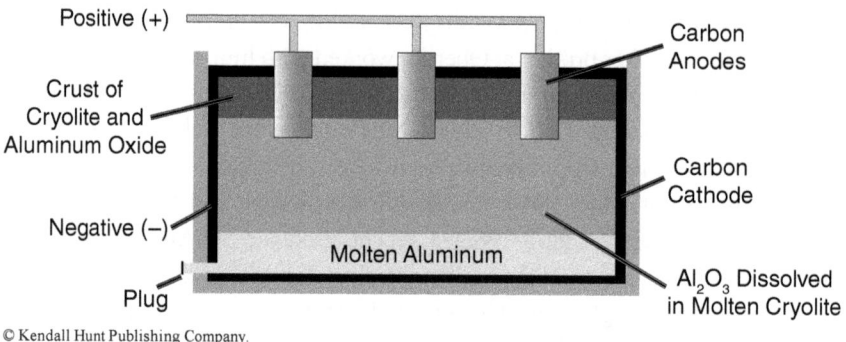

Figure 5   Hall's Electrolytic Process

Hall, described as private and secretive, never married nor did his sister. He loved playing classical music on the piano and attending the opera. He was a very religious person and left $10 million of his $30 million fortune to Oberlin College (established 1833) and $5 million to Berea College (established 1855) in Kentucky, both of which had religious roots and missions. Hall died in Daytona, Florida, in December 1914, four months after World War I began. A diseased spleen, an organ near the stomach and intestines, caused his death. Another source attributed his death to leukemia.

Paul Louis Héroult (1863–1914), a free-spirited, imaginative metallurgist born in Harcourt, France, the same year as Hall, invented the same aluminum process in April 1886, two months after Hall. Héroult spent a brief part of his childhood in England and in 1882 entered the School of Mines in Paris where he began investigating the electrolysis of aluminum. Héroult received a French patent on his process on 23 April 1886 and with the assistance of his new associate, Jules Dreyfus, first licensed it in 1887 to the Société Métallurgique Suisse (later Aluminum Suisse) in Neuhausen. He served as the Neuhausen plant's technical director. To expand aluminum's production, Héroult and Dreyfus obtained financial support from the Bank of Goldschmidt in Paris. They established the Société Electrométallurgique Français in 1888 and constructed a second plant in Froges in the Isere and a larger plant in 1893 at La Praz, Savoy, in the Maurienne Valley. Héroult's discovery resulted in the development of Europe's aluminum industry, licensing his patent in France and Germany; in England to the British Aluminum Company established in 1894 in Foyers, Scotland; and later in other European and global sites.

Héroult filed for a United States patent on 22 May 1886, but Hall established his priority with his February 1886 letters and won the patent disputes on the related electrolytic separation processes of Bunsen in 1854 and Sainte-Claire Deville in 1859. The United States Patent Office agreed that Hall's aluminum process was sufficiently different from all others. Héroult died of typhoid fever in April 1914, the same year as Hall, while on his yacht (which served as his home) in the Mediterranean Sea near Antibes, a city in southeast France.

Originally a precious and expensive metal aluminum found wide use because of its strength, lightness, and good electrical conductivity. Its main uses were in telephone lines, aircraft, and zeppelins. Because of commercial production, the price of aluminum dropped from $10 per pound in 1855, to $5 per pound in 1886, to 70¢ per pound in 1893, and 18¢ per pound in 1914. The Hall-Héroult process remains the main industrial method of producing aluminum.

PRICE OF ALUMINUM	
Year	*Price (per pound)*
1855	$10.00
1886	$ 5.00
1893	$ 0.70
1914	$ 0.18

ALUMINUM PRODUCTION IN THE UNITED STATES			
Year	Tons	Year	Tons
1883–92	280	1920	6,902
1883–1902	13,701	1921	27,266
1903	3,318	1922	36,816
1904	4,050	1923	64,329
1905	5,405	1924	75,282
1906	7,062	1925	70,058
1907	8,162	1926	3,693
1908	5,338	1927	81,803
1909	14,540	1928	105,272
1910	17,701	1929	113,986
1911	19,198	1930	114,518
1912	20,903	1931	88,772
1913	23,639	1932	52,444
1914	28,986	1933	52,562
1915	45,252	1934	37,088
1916	57,553	1935	59,647
1917	64,930	1936	112,464
1918	62,362	1937	146,360
1919	64,238	1938	143,441

Petroleum

The availability of natural resources established the history of energy consumption in the United States, dividing its energy history into three time spans: (1) the wood period, (2) the coal age, and (3) the petroleum era. The wood period began with the American Revolution and reached its peak in 1850 when wood supplied 90 percent of the energy consumed in the United States. Wood consumption decreased rather regularly after that, falling to 50 percent in 1885 and to only 5 percent in 2000.

The coal age began in the second half of the nineteenth century when coal consumption increased steadily. Coal supplied 10 percent of the total energy consumed in 1850, and by 1885 consumption rose to 50 percent. Coal and wood were at that time the United States' major energy sources. The use of wood as an energy source continued to decrease in the nineteenth century, whereas coal consumption attained a maximum in 1920 before starting to decline, dropping to approximately 45 percent of the total energy consumed in 1945 and to 20 percent in 2000.

Edwin Drake (1819–90) made the first significant petroleum discovery in 1859, but the petroleum era did not begin until the twentieth century. Petroleum, which provided 5 percent of the United States' energy demand in 1890, was with natural gas the source of 45 percent of energy consumed in 1945. Consumption in the United States and globally has increased steadily since 1945.

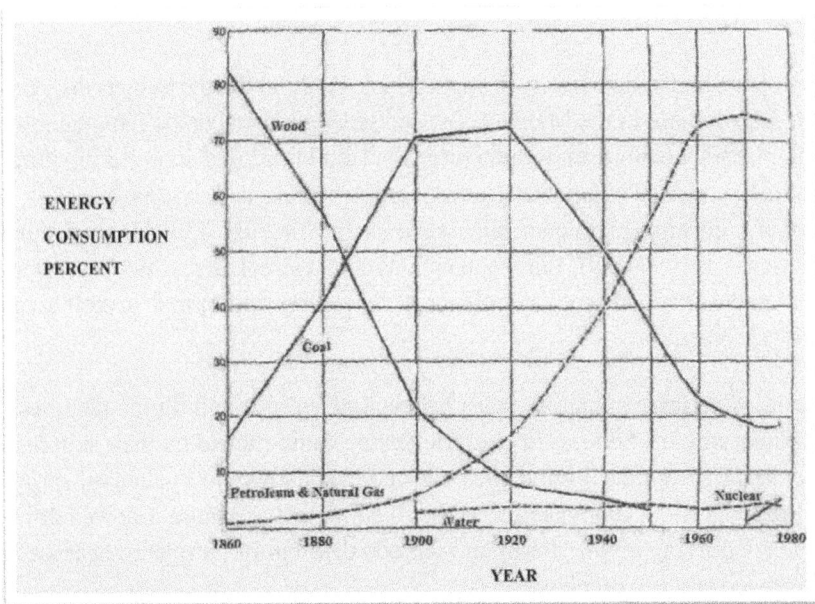

Figure 6   Energy Sources in the United States

Origin of petroleum, early use as a medicinal and as an illuminant

Eighteenth- and nineteenth-century scientists disagreed on petroleum's origin and formation. They debated whether petroleum had a vegetable, marine, or mineral origin and whether it resulted from gradual decomposition and fermentation or by distillation caused by subterranean heat. The Russian scientist Mikhail Lomonosov (1711–65) at St. Petersburg in 1757 proposed a biological origin (biogenic theory), whereas the German naturalist Alexander von Humboldt (1769–1859) at Berlin and the French chemist Louis Joseph Gay-Lussac (1778–1850) at Paris at the beginning of the nineteenth century, believed that petroleum was a primordial material or fundamental substance and that it had erupted from a great depth within Earth (abiogenic theory). The Russian chemist Dmitri Mendeleev (1834–1907) at St. Petersburg in 1877, also rejected petroleum's biological origin and agreed with Humboldt and Gay-Lussac on its origin. Mendeleev believed that metallic carbides found deep inside Earth reacted with water at high pressure to give acetylene gas which then underwent additional reactions to give heavier petroleum hydrocarbons. Since the 1920s most scientists have accepted the biogenic theory that petroleum originated from tiny plants and animals that died 10–20 million years ago and whose remains sank below Earth's swampy surface where they decayed and where heat and pressure transformed them into oil and gas.

Petroleum had a wide variety of applications. By 3000 BC the Sumerians, Assyrians, and Babylonians in the Middle East used seepages of asphaltic bitumen found near Hit in Mesopotamia in architecture, road building, and in waterproofing ships. Petroleum burned in lamps dates from ancient times, and in some societies, such as Persia's, burning petroleum released a sacred fire. By 750 AD, and during the Crusades of 1096–1291, European nations added Greek fire, which was naphtha the oily liquid distilled from petroleum at 80–110°C and stored in bottles, to their military arsenals.

In the Americas, Indian societies believed petroleum had divine qualities, which explains why the Senecas in western Pennsylvania rubbed on their skin petroleum that seeped from the ground and collected in shallow pools called oil springs. The U'wa Indians of Colombia decreed petroleum sacred because it flowed through the veins of a living mother Earth and forbade drilling for petroleum because drilling destroyed nature's lifeblood.

As early as 1526 numerous observers in the Americas, including Indians, missionaries, soldiers, and explorers, had described their discovery of oil springs, the most significant of which were those in Oil Creek in western Pennsylvania and first mentioned in 1767. By mid-nineteenth century petroleum from oil springs had achieved the status of a cure-all in medicine. An 1850s advertisement by Samuel Kier (1813–74), an early petroleum refiner in Pittsburgh, described petroleum (rock oil) as a natural remedy, useful for treating cholera, diarrhea, piles, rheumatism, gout, burns, pimples, and many other ailments. Kier began selling his distilled rock oil after he discovered that the medicinal oil his wife's doctor treated her with was petroleum. He sold his rock oil in half-pint (16 fluid ounces) bottles for $1.00, selling 240,000 bottles of rock oil by 1858.

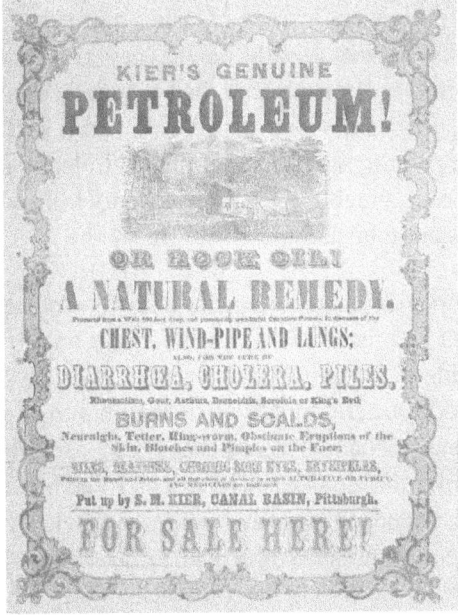

Reprinted with permission of *The Drake Well Museum, Pennsylvania Historical and Museum Commission*

Figure 7   Samuel Kier's Petroleum Advertisement

Petroleum's importance as an illuminant began in the 1850s. Whale oil burning lamps enjoyed wide use, but whale oil's high cost of $1.40 per gallon prompted the search for cheaper burning substitutes. The first substitute for whale oil was an oil that the Canadian surgeon and geologist Abraham Gesner (1797–1864) distilled from Nova Scotian coal in 1846. His distillations paralleled the experiments of James Young (1811–83), a Scottish technologist, who in 1847 distilled coal and oil shale at his Bathgate site and at other Scottish sites to obtain a crude oil. Young patented his process in 1850. Gesner, who left eastern Canada for New York City in 1853, called his product *keroselain*, from the Greek words for *oil* and *wax*, and patented it in 1854 with the name *kerosene*. His oil sold well but had a bad odor because it contained sulfur.

A major search for new sources of petroleum began on 30 December 1854 when Jonathan G. Eveleth (1821–61) and George H. Bissell (1821–84), two New York City law partners, established the Pennsylvania Rock Oil Company of New York, the first oil company in the United States. After noticing the chemical similarities between Gesner's coal oil illuminant and petroleum, Bissell proposed that they drill directly for petroleum instead of using petroleum by-products from salt water (brine) wells, as was usually done. They also heard that Samuel Booth (1810–88), a chemist at Pennsylvania, had distilled some crude petroleum that Kier obtained in 1849 from one of his salt water wells on the shore of the Allegheny River. Booth reported that the petroleum distillation yielded an illuminant (kerosene) suitable for lamps. Kier took Booth's advice and burned the kerosene fraction in his lamps.

The new use for petroleum required the discovery of new petroleum deposits. Bissell and Eveleth had received a favorable report on oil springs on the Hibbard Farm near Titusville in western Pennsylvania, and 17 days later on 30 November 1854 they purchased 100 acres for $5,000, with rights on an additional 12,000 acres. The crude petroleum obtained there burned reasonably brightly when tested in lamps, but it gave off much smoke and often a bad odor because of its sulfur content. Bissell and Eveleth intended to recruit investors to develop the Titusville oil springs, but to attract potential investors they had to determine the petroleum's quality. Benjamin Silliman, Jr. (1816–85), the Yale chemist they hired to make an analysis, distilled the crude petroleum and showed that it consisted of several lighter oil fractions of different densities, such as naphtha and kerosene, and heavier fractions, that included tars, lubricants, and waxes.

Silliman used a Bunsen photometer to compare the intensity of the light emitted by the fractions with that of other illuminants and found that the kerosene fraction had excellent illuminating properties. It was the most important fraction he separated, and because it burned cleanly kerosene soon replaced whale oil in lamps. Silliman's April 1855 *Report on the Rock Oil, or Petroleum, from Venango Co. Pennsylvania*, for which the Pennsylvania Rock Oil Company paid $526.08, suggested that high temperatures could crack (break) the heavier fractions to give higher yields of the lighter fractions, particularly the kerosene that Bissell and Eveleth wanted, and gasoline which had no use at that time.

Bunsen, an excellent experimental chemist, had invented a photometer called the grease spot photometer in 1843. It consisted of a horizontally-placed graduated measuring stick on which sat, at a right angle to the measuring stick, a white paper screen with a translucent grease spot at its center. A light source of known intensity (candle power) sat at a known distance on one side of the screen. Bunsen placed a light source of unknown intensity on the other side. If both sides received equal illumination the grease spot seemed to disappear. If one side of the paper screen received more light than the other, that side appeared brighter but had a dark spot in the center where the light passed through the grease spot. To measure the unknown intensity the operator moved the screen back and forth between the two light sources to find the position at which the grease spot disappeared. The intensity of illumination was now equal on both sides of the screen. The equation below enabled Bunsen to calculate the unknown candle power.

$$\text{candle power}_1/\text{distance}^2{}_1 = \text{candle power}_2/\text{distance}^2{}_2$$

First large petroleum discovery

Silliman's report on the fractional distillation of crude petroleum to give kerosene brought in a group of New Haven, Connecticut, investors that included Silliman. On 18 September 1855 the Pennsylvania Rock Oil Company of New York became the Pennsylvania Rock Oil Company of Connecticut, capitalized at $300,000. Bissell and Eveleth were the majority shareholders. Little activity ensued because of the 1857 panic and hard economic times until James Townsend (1825–1901), secretary, treasurer, and later president of the City Savings Bank of New Haven and president of the new company, hired Drake in 1857 to do the drilling and paid him an annual salary of $1,000.

Drake, to whom Townsend gave the title of colonel, was a retired railroad conductor formerly with the New York and New Haven Railroad Company now living in New Haven. He had no drilling experience but was available and had a free railroad pass that enabled him to travel to Titusville. Townsend then forced out Bissell and Eveleth and in March 1858 established the Seneca Oil Company of Connecticut as a new subsidiary of the Pennsylvania Rock Oil Company of Connecticut. He named Drake the general agent of Seneca Oil and leased the Titusville oil property to Seneca Oil.

To assist him in the drilling Drake hired William A. (Uncle Billy) Smith (d. 1890) for $2.50 a day. He was an experienced salt water well driller, blacksmith, and tool maker from Salina, Pennsylvania. After numerous delays and obstacles they constructed a derrick and began drilling near the end of May 1859. On 28 August after three months of drilling, they made the first large petroleum discovery at Titusville on the bank of Oil Creek at a depth of 69½ feet. The well produced 400 gallons in 24 hours. Before Drake's discovery Titusville produced 10 gallons total per day.

Drake, who transferred the techniques of boring for salt water to drilling for oil, made an important engineering advance still in use, when he introduced the pipe-driving method. He drove several 10-foot sections of a cast iron pipe, ½ inch thick and 3 inches in diameter, into the ground at the well's opening. It kept the petroleum flow line to the surface free from any cave-in of the surrounding clay and quicksand and prevented any underground water from flowing into and filling the hole. Drake and Smith used a 6-horsepower steam engine for drilling. It cost $500.00 at the factory, an expense few prospectors could afford at that time. Their tools weighed 100 pounds and cost $76.50.

Reprinted with permission of *The Drake Well Museum, Pennsylvania Historical and Museum Commission*

Figure 8   Drake's Derrick at Titusville

To store the 300 barrels of petroleum obtained from his first successful discovery, Drake collected as many empty 40-gallon wooden whiskey barrels as he could and built two large open vats, each holding 25 barrels.

Drake never patented his drilling innovation, likely because he could not prove originality and priority, and after 1866 lost all the money he made in the petroleum business in bad investments. He spent his remaining years in ill-health and with no source of income until the state of Pennsylvania awarded him a $1,500 annual pension beginning in 1873. His grave is in the Titusville cemetery.

Nineteenth-century methods of finding and drilling for petroleum

In the Oil Creek region of Pennsylvania, petroleum lay in shallow deposits, after having seeped through the soil, and required little searching to find. When petroleum wasn't seeping through the ground, prospectors searched for and found petroleum in porous rock (sandstone, a silicate rock) or trapped in pools between different strata. Petroleum deposits often lay near underground salt water deposits and presented a problem for salt miners because prior to 1855, the beginning of kerosene's widespread use as an illuminant, petroleum had less value than salt and had become more of a nuisance.

Diviners, spiritualists, and smellers in the 1860s claimed they could find petroleum. Diviners and spiritualists attributed their special powers with a divining rod (a forked twig of witch hazel or peach held vertically by the prongs) to their possession of an unusual positive charge of electricity. The electricity acted on the rod and directed the rod to petroleum deposits located below the surface. Their fees ranged from $25 to $100. Smellers claimed their acute sense of smell enabled them to discover petroleum deposits trapped several hundred feet underground.

Upon locating a petroleum deposit the two commonly used drilling methods were the jigging down spring pole method and the kicking down spring pole method. Both were manual methods and effective only for shallow drilling. In the more common jigging down method, two workers walked back and forth on a ground level, see-saw type of wooden horizontal platform located near one end of a 10–15 foot flexible horizontal wooden pole or walking beam. They set the wooden pole over a fulcrum and firmly secured it at the other end. A strong cable or rope attached to the platform extended vertically upward 5–8 feet where it wound tightly around the free end of the wooden pole. Another rope with the drill, a long 3-inch diameter iron rod with a steel edge attached, hung vertically near the free end of the wooden pole. Walking back and forth on the platform lowered and raised the platform which alternately lowered and raised the wooden pole and drill. A third worker guided the drill into the ground.

Copyright © Samuel T Pees. Reprinted by permission of the Petroleum History Institute.

Figure 9 Jigging Down Spring Pole Method of Drilling 1860

Kicking down was the faster of the two methods. Workers hung stirrups two or three feet from the fulcrum of a 10–15 foot flexible wooden pole, or a large sturdy branch from a tree trunk, secured to the ground at one end and extending horizontally at about a 45° angle. A strong rope with the drill, a long 3-inch diameter iron rod with a steel edge attached, hung from the wooden pole. Two or three workers placed one foot in each stirrup and by kicking outward and downward pushed the pole and drill downward and then allowed them to spring back. Experienced workers kicked down at the rate of two strokes a minute. A third worker guided the drill into the ground.

Reprinted by permission of the American Oil and Gas Historical Society

Figure 10   Kicking Down Spring Pole Method

Deeper oil drilling required abandoning the two older and simpler drilling methods and constructing a wooden derrick about 30–40 feet high and identical to those used in drilling for salt water. The derrick had a windlass (capstan) at the bottom for winding and unwinding the rope attached to the drill and a pulley at the top over which ran the rope. The rope, attached at one end to the windlass, passed over the pulley and had its other end attached to the drill. The horizontal pole set over a fulcrum, had its one end secured to the ground a short distance from the derrick and its free end attached firmly to the drill. The arrangement was similar to the attachment of the stirrups and the platform used in shallow manual drilling.

Three feet per day was a good average drilling rate. In soft slate workers drilled about 6 feet, in sand or rock conglomerate they drilled only about 2–3 inches. Flooding, cave-ins, and broken tools were common drilling problems. To drill to the usual depth of that time, about 200 feet, cost $1,000–$1,200. Steam-powered drills soon began to replace manual drilling. In 1865, a 12-horsepower steam engine cost $2,200, a 15-horsepower engine cost $2,500. The derrick or rig used previously in salt drilling and the steam engine now became an integral part of oil drilling equipment.

Copyright © Samuel T Pees. Reprinted by permission of the Petroleum History Institute.

Figure 11 Drilling with a Steam Engine

Petroleum storage and refining

Following Drake's 1859 discovery, prospectors from everywhere in the country rushed to Titusville, matching the enthusiasm of the California gold rush ten years earlier. A second well was operating in November 1859, a third well that produced 75–80 barrels per day was operating in March 1860. By summer 1860, 12 wells were operating, and by the end of 1860, 74 wells in the Oil Creek region were yielding 1,165 barrels per day. Pennsylvania remained a major petroleum producer, providing about 90 percent of the United States' total in 1885, but its production decreased from 30 million barrels in 1882 and 33 million barrels in 1895 to 19 million barrels in 1901, and 9.4 million barrels in 1908. After that petroleum production shifted to Ohio, Oklahoma, California, Wyoming, and Texas.

CHAPTER 2: The Technological Transformation   41

Reprinted with permission of *The Drake Well Museum, Pennsylvania Historical and Museum Commission*

Figure 12   Early Wooden Derrick and Petroleum Storage

In the early years of the petroleum industry, field workers pumped the petroleum from a petroleum deposit directly into large open vats and then transferred it to 40-gallon wooden whiskey barrels (since 1872 one barrel contains 42 gallons of petroleum). Wagons, ships, railroads in 1862, and pipelines in 1865–66 transported the petroleum around the country or to refineries.

Reprinted by permission of the American Oil and Gas Historical Society

Figure 13 Shipping Petroleum by Rail (late 1865)

Because the gasoline, kerosene, heavy oil, and other fractions that make up crude petroleum have different boiling points, distilling the petroleum in a small laboratory still, as Silliman and Booth did, or in a refinery, separated the fractions. The first refinery, constructed in Titusville in autumn 1860, cost $15,000. By the end of the year 15 refineries were operating, among them the refinery that the chemist Luther Atwood (1826–68) of the Queen's County Oil Works in Long Island, New York, constructed to crack crude petroleum and heavy coal oil distillates into fuel oils using superheated steam. The number of refineries increased to 60 in 1863 and to 90 in 1865.

A small refinery distilled about five barrels of petroleum per day and cost about $200 to construct. A larger 10-barrel per day refinery cost $300. One barrel of distilled petroleum yielded 60–65 percent kerosene (bp 180–300° C), 10 percent gasoline (bp 220°C), 5–10 percent naphtha (petroleum products with boiling point less than 240°C), and the remainder tar and residue. By this time major refining centers emerged in Cleveland with 12,732 barrels per day capacity, New York City area with 9,790 barrels per day capacity, Pittsburgh with 3,500 barrels per day capacity, and Baltimore with 1,098 barrels per day capacity.

Harold F. Williamson and Arnold R. Daum (1959). *The American Petroleum Industry, The Age of Illumination 1859-1899*. Northwestern University Press.

Figure 14 A Petroleum Refinery

Oil refining and distillation practices were generally primitive. In fact, fraudulent or poor distillation may have contributed to the Chicago fire of 1871. The lantern that Mrs. O'Leary's cow allegedly kicked over to start the fire, which killed 300 people and resulted in $200 million property loss, may have contained highly combustible gasoline mixed with kerosene and other combustible petroleum fractions. Kerosene remained the top selling petroleum fraction until 1911, but with the mass production of automobiles, after 1911 gasoline consumed in the internal combustion engines of automobiles became the petroleum fraction in greatest demand.

One of the heavier petroleum fractions found use as a lubricant for horses' hoofs and harnesses, for wagons, and in the railroad industry. Elijah McCoy (1843–1929), a Canadian-born, African-American inventor educated in Edinburgh, Scotland, and living in Detroit, Michigan, established the Elijah McCoy Manufacturing Company to manufacture lubricants. From 1872 to 1876 he obtained the first patents on petroleum lubricants for use in steam engines and machinery. McCoy's automatic lubricator, or lubricator cup, allowed small drops of oil to fall continuously on the moving parts of a machine, making it no longer necessary to stop the machine and apply the lubricant manually. Petroleum now replaced tallow, a fat obtained from sheep, cows, and other animals, as the chief source of lubricants. McCoy's later patents, dating from 1892 to 1926, gave him a total of 44 patents.

Petroleum discoveries after Titusville

The first major American petroleum discovery after Titusville occurred in 1896 at Corsicana, Texas, where 300 wells produced 800,000 barrels over four years. Five years later, on 10 January 1901, Captain Anthony F. Lucas, born Antonio Francisco Luchich (1855–1921), made a more sensational petroleum discovery at Spindletop, Texas. Lucas, a graduate of the Polytechnic Institute in Graz, Austria, and a student at the Austrian Naval Academy, emigrated to the United States in 1879.

At Spindletop, a town of 9,400 near Beaumont in southeast Texas, Lucas No. 1 oil well spouted a gusher 200 feet high. The petroleum deposit lying 1,100 feet below the 140-acre Spindletop mound produced 900,000 barrels in its first nine days and 3.5 million barrels in its first year. The oil glut dropped the price of oil as low as 3¢ per barrel in 1901. Spindletop produced 17 million barrels in 1902, helped by the first application of rotary drilling at Spindletop in 1901. By 1902 major oil companies, such as Mobile, Sun, Gulf, Texaco, and Humble, were operating at Spindletop, which still produces 600 barrels per day.

Other discoveries included Los Angeles in 1890, Mexico in 1910, Oklahoma in 1927, and the East Texas Kilgore Longview region in September–October 1930 and early 1931. The 200-acre East Texas field was 42 miles long and four to eight miles wide. Columbus Marion "Dad" Joiner (1860–1947), a crafty lawyer and veteran wildcatter, made the first significant East Texas strike, 300 barrels per day, in Rusk County after three unsuccessful years of drilling. The legendary Texas oilmen W. A. "Monty" Moncrief (1895–1986), H. L. Hunt (1889–1974), and Sid Richardson (1891–1951) made fortunes in the East Texas oilfield.

CHAPTER 2: The Technological Transformation

UNITED STATES PETROLEUM PRODUCTION AND PRICE	
Production (barrels per year)	*Price* (per barrel)
1859: 2,000	1859: ($20.00) 40¢
1860: 500,000	1860: $2.00 ($9.60)
1861: 2.01 million	1861: 49¢
1862: 3.05 million	1862: $1.05
1869: 4.00 million	1869: $5.64
1879: 20.0 million	1879: 86¢
1884: 24.2 million	1884: 85¢
1889: 35.0 million	1889: 77¢
1890: 45.8 million	1890: 70¢
1899: 57 million	1899: $1.13
1901: 69 million	1901: 96¢
1924: 700 million or about 70 percent of the world's total production	1924: $1.43

Thermal and high-pressure catalytic cracking of crude petroleum

In 1909–10 William Burton (1865–1934) and his associates Robert Humphreys and F. M. Rogers at Standard Oil Company, Indiana, began experiments that led to commercial-scale, high-pressure, 60–75 pounds per square inch (psi), thermal cracking of petroleum. Their process in which Burton cracked gas oil in a 50-gallon still to obtain 20–25 percent gasoline. doubled the yield of gasoline. The use of high pressures in industrial chemistry was a new technique that the German chemical company BASF in Ludwigshafen first used in 1909 for the high-pressure synthesis of ammonia (NH_3) from the gaseous elements hydrogen and nitrogen.

Standard completed construction of its first commercial-scale still in November 1912, and in January 1913 the first battery of stills began batch operation at 75 psi. Each still consisted of a steel-plated horizontal drum 8 feet in diameter, 30 feet long, and had a capacity of 8,250 gallons of gas oil. Gasoline yields averaged about 25 percent. Standard constructed another 240 stills in July 1913. Burton served as Standard Indiana's president from 1918 to 1927. His process remains widely used today.

Almer McAfee (1886–1972) at Gulf Oil in Port Arthur, Texas, in 1915 patented the first catalytic high- pressure cracking process, using aluminum chloride (alchlor process) as the catalyst and a pressure of 60–100 psi. He developed an improved low-cost, commercial-scale process in 1923, and by the mid 1920s Gulf constructed twenty-seven 27,000-barrel cracking stills at Port Arthur, and three more at Fort Worth, Texas, to produce No-Nox gasoline. Eugene Houdry (1892–1962), a French mechanical engineer in Beauchamp near Paris, developed a second catalytic cracking process in 1927 using clay-type silica-alumina catalysts. His process doubled the yield of high-quality fuel obtained from crude petroleum.

Houdry moved to the United States in 1930, and that same year his company, the Houdry Process Corporation in Paulsboro, New Jersey, joined with Vacuum Oil Company in Paulsboro to develop his catalytic cracking process. Following Vacuum Oil's merger with Socony (Standard Oil of New York) Houdry tested his process in 1936 in Socony-Vacuum's 2,000-barrel per day Paulsboro refinery. The next year, in 1937, Houdry reached an agreement with Arthur E. Pew (1899–1965), vice-president of Sun Oil, to use the process on a commercial scale at its $450,000 Marcus Hook 15,000-barrel per day Pennsylvania refinery. The yield was 48 percent of 81-octane gasoline.

United States petroleum reserves

In addition to its significant coal deposits and then-uncertain petroleum sources, the United States possessed considerable oil shale at sites in Colorado and Utah. At the turn of the twentieth century, with petroleum consumption and fear of future shortages steadily increasing, David T. Day (1859–1925), a chemist-geologist and chief of the Mineral Resources Division of the United States Geological Survey from 1886 to 1907, began examining the oil shale sites. The Geological Survey considered oil shale distillation a good source of petroleum and encouraged its development, especially after Day's 1908 report pointed out that the United States lacked sufficient petroleum reserves to meet its future needs and would exhaust its mid-continent fields by the end of 1923.

Several other pre-World War I reports in 1909 and 1914 made the same gloomy prediction about future petroleum supplies in the United States. As a result the Bureau of Mines, established in 1910 as a branch of the Interior Department, in 1913 began a program of deriving oil from shale, which estimates in the 1920s put at 92–105 million barrels of oil with an extraction cost of $1.00 per barrel compared to petroleum's production cost of $2.50–3.00 per barrel. The discovery of the Oklahoma oil fields in 1927 and the huge East Texas oil fields in 1930–31 ended the gloom at least until after World War II.

At that time the Bureau of Mines initiated coal conversion to liquid fuel (synthetic liquid fuel) processes and resumed research on oil shale. Falling petroleum prices and political issues ended the Bureau's short-lived program by the mid-1950s. The oil crises of 1973–74 and 1979–81 led to resumed research in synthetic liquid fuel processes. Research activity ended in the 1980s for the same reasons as the earlier programs, economics and politics.

The United States in the 2000s consumes about 19 million barrels of petroleum per day, about 50 percent of which is gasoline. It imports about 7 million barrels per day, or 37 percent of its total consumption. It remains the world's largest consumer, far outstripping its nearest competitors, China at 6.5 million barrels per day, and Japan at 5.6 million barrels per day. The United States has 3 percent of the known global reserves but consumes 25 percent of the 87.5 million barrels per day produced globally in 2011. The imbalance between domestic production and consumption has persisted since the 1960s although a 1980s–90s advance in petroleum recovery has revolutionized domestic petroleum and natural gas production and reduced the gap between production and consumption. The advance, the combining of horizontal drilling and hydraulic fracturing, which began in 1986–91 in the Barnett Shale located in the Dallas-Fort Worth region of Texas, has enabled the United States to recover petroleum and natural gas from its extensive but deep, dense shale rock deposits. The fast growing application of the combined process to other shale deposits in North Dakota (Bakken Shale), Pennsylvania (Marcellus and Utica Shale), New York (Marcellus and Utica Shale), West Virginia (Marcellus Shale), Texas (Barnett and Eagle Ford Shale), and other states has reduced significantly the United States' reliance on petroleum imports to meet its domestic demand.

Horizontal drilling and hydraulic fracturing have long histories. Horizontal drilling, previously called directional drilling, dates from 1934 in Conroe, Texas. It remained insufficiently developed until 1973 but has dominated drilling in the 2000s. Of the 1,756 rigs operating in 2013 in the United States, 1,100 were horizontal drilling rigs.

Hydraulic fracturing (fracking), which began in 1947–49 in Oklahoma and Texas, recovers oil and gas trapped in dense rock (shale) formations by blasting a mix of water (90 percent) to fracture the shale, sand (about 9 percent) to keep the fractures open, and chemicals (acids, salts, alcohols, gels, about 1 percent) at high pressure into the formation to release the previously unrecoverable trapped oil and gas. George Mitchell (1919–2013) in the Woodlands, Texas, founder of Mitchell Energy and Development Corporation (established in 1946 and 1949 under various names and sold for $3.5 billion in 2002 to Devon Energy in Oklahoma City), pioneered in hydraulic fracturing from 1946, developing new techniques in 1986 and 1997. He combined his new-improved fracking process in which he invested ten years and $6 million with the advances in horizontal drilling and successfully extracted the natural gas trapped in the Barnett Shale formation. Mitchell's revolutionary technological innovation made the recovery of natural gas and petroleum from shale rock economically possible in the United States (and globally). His innovation accounted for the significant increase (20 to 30 percent from 2008 to 2012) in US domestic production of oil and natural gas that followed.

NOTABLE DATES IN PETROLEUM HISTORY	
Date Established	*Institution*
1849	Department of Interior
1879	Geological Survey
1910	Bureau of Mines
1919	American Petroleum Institute

IMPORTANT EVENTS IN THE TECHNICAL DEVELOPMENT OF THE AMERICAN PETROLEUM INDUSTRY	
Year	Event
1901	The brothers Al and Curt Hamill, well-known East Texas drilling experts, made one of the first uses of the newly-invented rotary drilling method in the Spindletop field near Beaumont.
1903	I. L. Dunn discovered the principle of using compressed air or natural gas injection for the secondary recovery of petroleum while working in Ohio's Macksburg pool. This is the Smith-Dunn or Marietta process.
1907	Petroleum workers near Bradford, Pennsylvania, accidentally discovered the principle that water flooding increased the recovery of petroleum from a pool.
1908	Petroleum companies introduced steel rigs and derricks of structural and tubular steel.
1911	I. L. Dunn and the Cumberland Oil Company successfully used compressed air injection for secondary recovery near Chesterhill, Ohio.
1912	Geologists started to map outcrops to locate favorable places for drilling wildcat wells.
1913	Engineers discovered accidentally that dynamite extinguished a flaming well.
1914	The rotary method succeeded in drilling softer formations not exceeding a depth of 3,000 feet.
1923	Rotary drilling became the most common drilling method. Engineers seriously began to apply sub-surface geology techniques as they developed drilling to a high degree of efficiency. They used the magnetometer, introduced from Germany, for geophysical exploration.
1924	Henry L. Doherty (1870–1939), a utilities expert who founded Cities Service Company in 1910, advanced the theory that petroleum in an undisturbed pool differed in character and behavior from petroleum at the surface. This led to new ideas and techniques of petroleum recovery.
1925	Introduction of micropaleontology and mineralogical techniques to make correlations in sub-surface geology began.
1925	The deepest well drilled reached about 7,000 feet.
1926	Petroleum companies developed the drill-stem tester for checking the fluid content of a petroleum formation without setting the casing.

1929	The development of the reflection seismic method extended the depth of mapping by many thousands of feet and detailing of structures.
1932	Petroleum companies drilling in Michigan's fields introduced the use of acid to stimulate petroleum production.
1933	Engineers introduced electric logging in the United States and used it in many petroleum fields to check, confirm, and modify correlations based on ordinary well logs.
1934	Horizontal drilling first done in Conroe, Texas.
1941	Radioactive well logging on a commercial basis began.
1949	Engineers successfully employed hydraulic fracturing of petroleum formations to stimulate production and increase recovery in Oklahoma and Texas.
1950	The combination of acidifying and fracturing petroleum formations stimulates production and increases recovery.
1954	The theory of combustion drive attracted much attention in the industry as a possible means of increasing petroleum recovery, especially of viscous oils or oil from old pools. Engineers successfully tested a dual-zone pumping installation in the Prentice Field in West Texas.
1955	Drilling in Louisiana reached a record depth of 22,529 feet with possible petroleum production at 21,605 feet.
1970	The age of the computer and complete automation of drilling systems began.
1986–91	Combined horizontal drilling and hydraulic fracturing began in the Barnett Shale in Texas.

Communications industry

Telephone

Alexander Graham Bell (1847–1922) grew up in a family in which his father and grandfather for years studied sound and methods of teaching speech to the deaf, including his almost deaf mother who began to lose her hearing when Bell was only twelve. Because of family health problems, the death of Alexander's younger brother of tuberculosis in 1867, the Bells emigrated from Edinburgh, Scotland, to Brantford, Ontario, Canada, in 1870 and then to Boston in 1872. In October 1872 Bell opened a school he named Vocal Physiology and Mechanics of Speech, for teachers of the deaf and to provide instruction to the deaf. His future wife, Mabel

Hubbard (1857–1923), who became deaf at age four after an attack of scarlet fever, came to him for instruction in 1873. That same year Bell became a professor of vocal physiology at Boston University and transferred his speech classes to the university.

Bell in 1874 met Thomas A. Watson (1854–1934), an electrician who worked in the Boston electrical shop of Charles Williams, Jr. In early 1875 they began working on Bell's design of a harmonic telegraph. Bell and Watson wanted to send from the harmonic telegraph (the transmitter) in an electric circuit, several tones (frequencies), as in a musical chord or the human voice, over a single wire at the same time and to have a receiver at the other end of the electric circuit vibrate at the frequencies of the transmitted sounds. At first they tested vibrating tuning forks of different frequencies which the German physicist Hermann von Helmholtz (1821–94) already had tried. Upon obtaining unsatisfactory results, they tested the vibrations produced by magnetized steel organ reeds.

Success arrived on 2 June 1875. Bell was at the receiver end of the electrical circuit. Watson was at the transmitter end in a separate room, tuning a reed of the harmonic telegraph by tightening it to vibrate at a different frequency, similar to tightening each string on a guitar to vibrate at a specific frequency. Watson had screwed the reed so tightly that it locked to the pole of the harmonic telegraph's electromagnet. When he plucked the reed to free it, Bell, at the other end of the circuit heard a *twang* in his telegraph receiver that sounded quite different from the usual whine that Watson's vibrating transmitter sent out. He heard the distinctive twang of a plucked reed, a sound with tones and overtones, coming to him over the wire. Bell and Watson concluded correctly that the tightly-screwed reed did not send an intermittent electric current as in an ordinary telegraph, but had acted as the transmitter diaphragm and sent an induced, varying (undulating) electric current through the circuit.

Source: Museum Victoria. Reprinted by permission

Figure 15 Bell's Vibrating Reed Used for a Receiver or Transmitter in a Harmonic Telegraph Circuit 1875

After much thinking and experimenting, during which time Bell drew a diagram of a human ear to understand better how humans receive sound vibrations, Bell finally had an explanation for the telephone's transmission of sound. The varying vibrational frequencies of a reed (or diaphragm) in the electromagnetic field surrounding the pole of an electromagnet generated an electric current that varied in intensity, or undulated, according to Faraday's principle of electromagnetic induction. Upon speaking into the telephone, the varying frequencies of the human voice caused the transmitting reed to vibrate at these same frequencies. The variable current passing through the electrical circuit (their telephone line) then vibrated the receiving reed (diaphragm) at the other end of the circuit at the same frequencies, which in turn caused the density of the air near the receiving reed to vary accordingly and the longitudinal sound waves to repeat the same sounds. The receiving reed pressed against Bell's ear had acted as the receiver diaphragm. Most important of all, the induced current proved strong enough to vibrate the reeds. After about an hour of plucking reeds and listening to the transmitted sounds, Bell gave Watson instructions for making the first Bell one-way telephone, and on the next day the primitive instrument transmitted recognizable but weak sounds of Bell's voice, not words, to Watson. Bell was a good piano player, and his knowledge of harmonics pitches, tones, and frequencies helped to clarify his thinking about vibrating reeds and variable frequencies.

Bell and Watson experimented on improving the harmonic telegraph (telephone) all summer, and in September 1875 Bell began to write specifications for his first telephone patent. His telephone had a major weakness. It required a stronger undulating current in the electrical circuit than electromagnets and electromagnetic induction provided. Bell knew in February 1876 that Elisha Gray (1835–1901), a Chicago inventor, had used a liquid water transmitter in his telephone system. Bell therefore tried to improve the voice transmission in his telephone by replacing the double electromagnet of his electromagnetic induction transmitter with a liquid water transmitter. In a liquid water system the resistance to a battery's current increased and decreased as the transmitter diaphragm vibrated. Bell's 1876 telephone patent specifications described a liquid transmitter that had a short wire attached directly to the diaphragm. As the diaphragm vibrated, the wire moved up and down in a liquid mercury conductor or in some other liquid conductor. When the wire went deeper, resistance decreased; as it rose again, resistance increased, so that a current flowing through the wire and liquid conductor varied as the frequency of the sound waves varied. Bell never mentioned explicitly a liquid water transmitter in his 1876 patent nor tested a liquid mercury transmitter as he did a water transmitter. He likely realized mercury's impracticality, and even if tested, mercury, a liquid metallic element and a good electrical conductor, would not provide enough resistance to produce a varying current.

Bell's financial backer and future father-in-law Boston attorney Gardiner Greene Hubbard (1822–97) had his lawyer, Marcellus Bailey (1840–1921) in Washington, DC, file Bell's patent application on 14 February 1876. Bailey filed only a few hours before Gray's attorney, William Baldwin, filed Gray's patent caveat on an almost identical liquid water telephone transmitter that Gray called a vocal telegraphical transmitter. Bell received his first telephone patent, US Patent 174,465, on 7 March 1876, calling it vocal telegraphic transmission.

On 10 March 1876, three days after receiving his first telephone patent, Bell's telephone carried its first intelligible sentence. Bell and Watson were on the top floor of a boarding house at 5 Exeter Place in Boston where Bell had rented rooms to secure greater privacy than the Williams shop afforded. To test the liquid water transmitter, Watson had gone to the other end of the electrical circuit, in Bell's bedroom one floor down. When he put his ear to the receiving telephone almost immediately he heard Bell's voice saying, "Mr. Watson come here, I want (to see) you!" Watson hurried to Bell's room, explaining that he heard faintly but clearly Bell's every word.

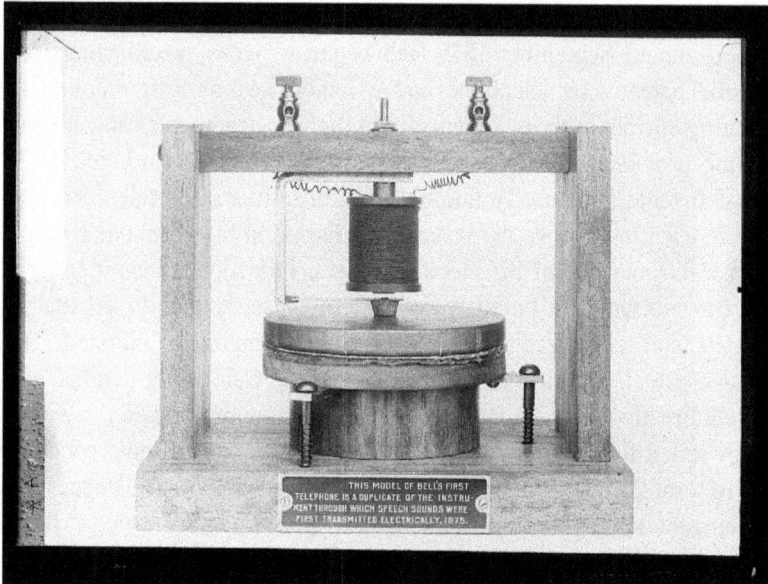

Courtesy of the Library of Congress

Figure 16 Bell's First Telephone—Harmonic Telegraph 1875

Bell, nevertheless, recognized the impracticality of a liquid transmitter, and in March 1876 he had resumed experiments on his electromagnetic transmitter. He improved its transmitting quality by replacing its double electromagnet either with a permanent iron or steel bar magnet that had a small coil of fine iron wire wound around and close to but not touching, one pole of the magnet; or with an iron or steel bar magnetized by an electrical current from a battery. He also introduced a thin iron disc mouthpiece that functioned as both the diaphragm and armature. The receiver at the other end of the circuit consisted of the same apparatus. Bell patented his electromagnetic transmitter in his second telephone patent, US Patent 186,787 of 30 January 1877, and in that patent changed the name of his transmitting system from vocal telegraphic transmission to electric telephone and telephone.

Bell demonstrated his improved electromagnetic telephone before the American Academy of Arts and Sciences in Boston in May 1876 and at the Centennial Exposition in Philadelphia in June 1876. It impressed participants, such as Brazil's long-ruling Emperor Dom Pedro II (1825–91) who had seen Bell's demonstration in Boston the previous week. On 4 October 1876 Bell and Watson held a conversation over the two-mile distance from Boston to Cambridge. The following year Bell introduced the telephone to England, giving a demonstration to Queen Victoria.

The French government in 1880 awarded him the Volta Prize of 50,000 francs ($10,000) for inventing the telephone.

In autumn 1876 before Bell fully grasped the telephone's potential, he and his two financial backers, Thomas Sanders (1839–1911) a successful Salem leather merchant, and Hubbard, offered Western Union Telegraph Company all rights to the telephone for $100,000. Ezra Cornell (1807–74) and his partners who founded Western Union in 1856 considered the telephone a poor buy, or as William Orton (1826–78), Western Union's president said, an electric toy. They declined Bell's offer, only to regret their decision once they recognized the telephone's significant potential and profitability. Western Union in March 1878 established its own telephone company, the American Speaking-Telephone Company with $300,000 capitalization, supported the claims of Bell's rivals, Gray and Thomas Edison, and challenged Bell's claim of invention.

Bell's controversy with Gray began in February 1877, only days after receiving his second telephone patent. One year earlier on the same day, 14 February, only a few hours after Bailey, Hubbard's attorney filed the first patent on Bell's telephone, Baldwin, Gray's attorney filed a caveat, a statement similar to a patent application, indicating that he was working on a vocal telegraphic transmitter (an electric speaking telephone). Gray's caveat described a telephone transmitter with a liquid having a high resistance such as water. It transmitted musical tones not the human voice and closely resembled Bell's patent on a mercury or other liquid conductor. Baldwin, however, advised Gray not to try to patent his telephone because the Patent Office already had notarized Bell's patent application before Gray filed his caveat. Bell did acknowledge in a letter of 2 March 1877 that he learned of Gray's caveat during a visit to the Patent Office in Washington on 24 February 1876. Later, in a court proceeding in April 1879, Bell said he discussed Gray's confidential patent caveat with Zenas Fisk Wilber, a patent examiner in the Washington Patent Office. Despite Bell's revelations, Gray conceded Bell's priority at the 1879 court proceeding, thereby ending their dispute. This was the first of over 600 patent claims and challenges Bell faced during the next 18 years.

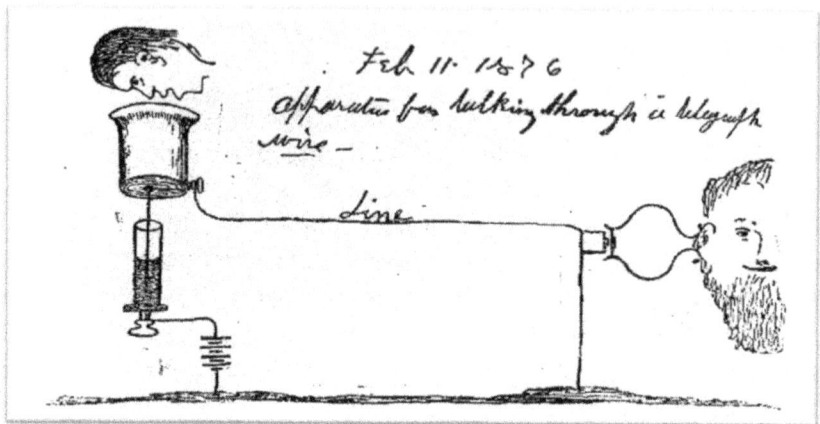

Figure 17 Gray's Patent Caveat

The dispute with Edison and Western Union occurred in February 1878. Edison had invented a telephone transmitter, or carbon button variable resistor transmitter (microphone), for which he filed a patent on 27 April 1877 but received it only on 3 May 1892. Edison's transmitter, in which the conductivity of carbon granules in a carbon box varied with the pressure the human voice acting on the diaphragm applied to the granules, was superior to Bell's and made the telephone practical.

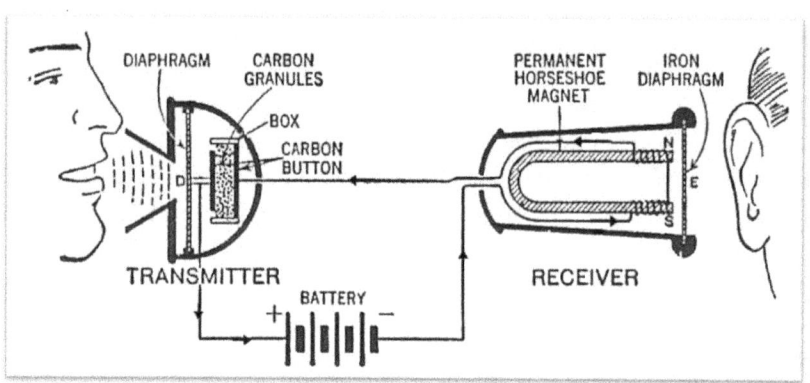

From *Visualized Physics* by Alexander Taffel. Copyright © 1959 by William H. Sadlier, Inc. Reprinted by permission.

Figure 18 Bell-Edison Telephone

The same principle of electromagnetic induction that Bell had applied in his first transmitter enabled Edison's carbon button transmitter and the telephone to function. The amount of current passing through the telephone's electric circuit varies with the frequency of the sound waves from the voice that strike the telephone transmitter's thin metal diaphragm and cause the diaphragm to vibrate or move back and forth. This either compresses the carbon granules in the carbon box or allows them to spread apart, thereby varying the electrical resistance and the amount of current that flows through the circuit. The varying current varies the strength of the electromagnet and the attraction it exerts on the iron diaphragm in the receiver, causing it to vibrate at the same frequency as the diaphragm in the transmitter and, therefore, of the speaker's voice. The current in the telephone circuit varies, the circuit does not open and close as in a telegraph. The telephone converts sound energy to electrical energy for transmission through the telephone lines and then converts the electrical energy to sound energy at the other end of the telephone line.

Despite the superiority of Edison's carbon box transmitter Edison could not circumvent Bell's telephone patent, and on 10 November 1879 Western Union gave up its telephone rights in return for 20 percent of the telephone rental income for 17 years, the life of the patent. Bell won that dispute, but he lost the fight over the telephone's greeting call. Bell, the sailor, wanted to use the nautical term *ahoy* as the greeting call, whereas around 1877, Edison suggested *hello* which came from *halloo* the traditional hunting call that Edison used to test his phonograph. Edison is not the name of any scientific unit, but Bell, spelled bel, is the standard unit for the intensity of sound. More commonly expressed as a decibel (dB), 1/10 of a bel, 0 decibels is the weakest sound anyone with normal hearing can hear, 20 decibels is a whisper, 60 decibels is normal conversation, 85 decibels will damage the inner ear and cause hearing loss.

To manufacture and promote telephone use Bell, Sanders, Hubbard, Watson, and three other Bell family members founded the Bell Company in July 1877. They reorganized and renamed their company the Bell Telephone Company in July 1878, the National Bell Company in March 1879, and the American Bell Telephone Company in 1880. One of the company's first policies, which Hubbard instituted and which remained in effect until the 1960s–70s was its telephone rental policy. No one owned a telephone, users had to rent one from the Bell Company at an annual cost of $20 for two telephones for two residences and $40 for two businesses. If desired the Bell Company would construct the line connecting the two telephones for $100–$150 per mile, or renters could construct their own lines.

After a rough beginning the American telephone industry grew rapidly. New Haven, Connecticut, had the first switchboard in 1878 with 50 subscribers. By 1880, 1,885 towns had telephone networks. In 1892 the United States had 240,000 telephones, 800,000 telephones in 1900, and about 3,132,000 telephones in 1907.

Long distance communication was the next challenge. On 3 March 1885 the American Bell Telephone Company established American Telephone and Telegraph (AT&T) as a holding company in New York City with $100,000 capital to build and operate the long-distance service that it initiated that year. Its first line connected New York City and Philadelphia and had a capacity of one-call. A second line with the same one call capacity connected New York City and Chicago. Users paid $9.00 for a five minute call.

Long distance communication throughout the United States became a reality beginning in 1894 when Michael Pupin (1858–1935) at Columbia placed toroidal induction coils in telephone lines. He showed that coils inserted about every 4–5 miles in overhead wires and every 1–2 miles in underground cables greatly reduced attenuation and distortion. Bell bought Pupin's patents in 1901 for $435,000, greatly expanding the telephone's application beyond the maximum of about 500 miles.

AT&T's growth continued in the decades following its establishment. In 1899 it absorbed the assets of its parent company, American Bell Telephone Company, and in 1908 it gained control of Western Union Telegraph Company by purchasing 30 percent of its stock. A 1913 agreement with the Department of Justice known as the Kingsbury Agreement forced American Bell to divest Western Union because the merger violated federal antitrust laws.

On 25 January 1915 Bell and Watson carried out the first transcontinental communication. With Bell in New York City, Watson in San Francisco, and Theodore Vail (1846–1926), AT&T's president, in Jekyll Island, Georgia, Bell repeated his original message of March 1876: "Mr. Watson come here, I want (to see) you." Watson replied that he would be glad to come, but it would take him more than five days to get there. The call took 23 minutes to go through and cost $20.70. Workers constructing the line installed 130,000 telephone poles and strung 13,800 miles of wire, completing the transcontinental line in five years on 17 June 1914 near the town of Wendover on the Utah-Nevada border.

CHAPTER 2: The Technological Transformation 59

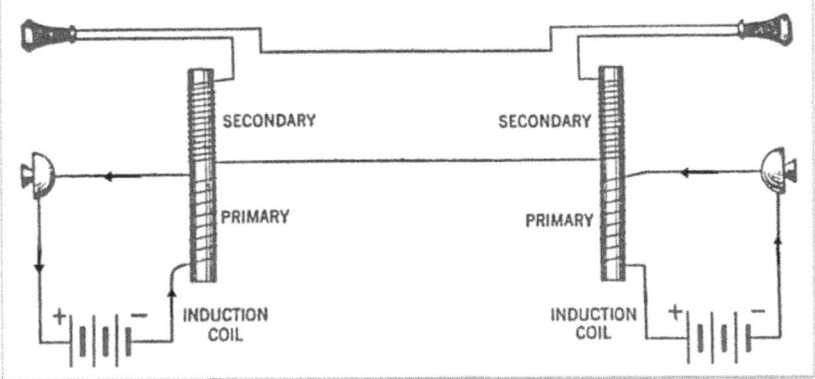

From *Visualized Physics* by Alexander Taffel. Copyright © 1959 by William H. Sadlier, Inc. Reprinted by permission.

Figure 19   Two-Way Telephone Circuit

While conducting experiments on the transmission of sound Bell devised an electric circuit consisting of an induction balance that canceled any interference produced by electrical currents in a telephone line. He also noticed that metallic objects near the balance disrupted it and caused a hum in the line. On 2 July, 1881 news broke that Charles Guiteau (1841–82), an evangelical attorney, disappointed office seeker, and tireless self-promoter, had shot the newly-elected Republican president, James Garfield (1831–81), as he was about to board a train at the Baltimore and Potomac Railroad Station in Washington, DC. Two bullets struck Garfield, hitting him in the arm and in the back. When Bell heard that the president's physicians failed to locate the bullet in Garfield's back, Bell, with the physicians' approval, applied his wand-shaped induction balance metal detector in an effort to save Garfield's life.

Bell's induction balance consisted of two coils of insulated wire, a battery, an interrupter (circuit breaker), and his telephone. He connected the ends of one coil to the battery and to the interrupter and the ends of the other coil to the telephone's posts. When Bell placed a piece of metal near the interrupter he heard a hum in the telephone receiver. When he moved the metal away the hum became fainter and faded completely when he moved the metal about five inches away. Bell successfully tested his balance, detecting small metal objects located in a variety of places, including inside the bodies of Civil War veterans. When tested on a conscious but wary Garfield it failed in two tests, on 26 July and 1 August. Bell's metal detector produced a continuous hum. The usually given explanation is that

Garfield lay on a new mattress design, one containing steel coil springs, and which neither Bell nor any of Garfield's physicians knew were present.

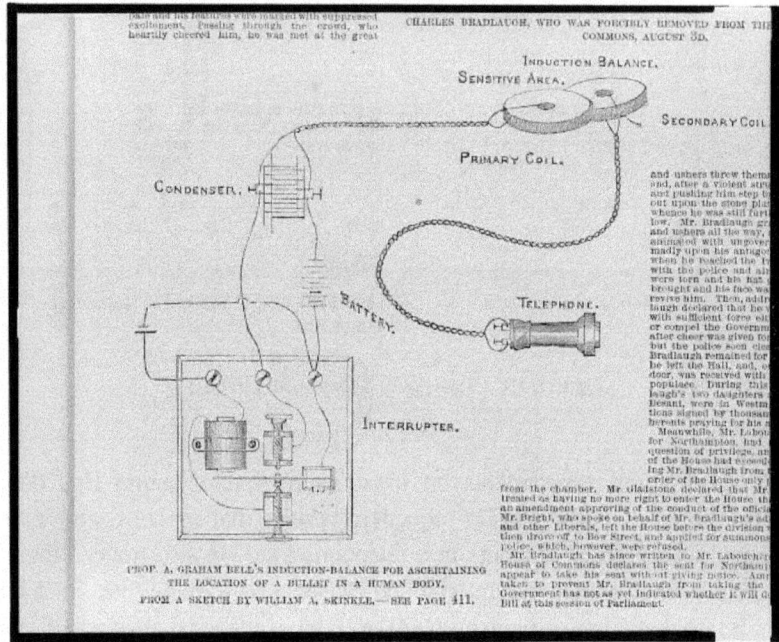

Courtesy of the Library of Congress

Figure 20 Bell's Induction Balance

Bell returned to his laboratory in Boston to improve the induction balance, and in the process invented the telephonic needle probe. The instrument consisted of two electrodes and a telephone receiver connected in a series circuit. One electrode was a flat metal plate, the other was a needle probe protected by shellac insulation except at its tip. Contact between the needle and metal produced an audible click in the telephone receiver. Bell, however, never tested the probe on Garfield whose weight had fallen from 200 to 120 pounds and was already near death with a severe infection caused by the fisted probing of the examining physicians' unsterilized hands.

Garfield died on 19 September. Guiteau, who pleaded insanity in his six-month trial, died by hanging at the District of Columbia jail on 30 June 1882. Bell's probe, at that time the only apparatus available for examining the body internally, eventually proved successful. Physicians employed his probe in locating bullets for removal from the wounded during the Sino (Chinese)-Japanese War (1894–95), the Boer War (1899–1902), and World War I (1914–18).

In addition to his inventions Bell and Hubbard financially supported publication of the journal *Science* beginning in 1882 and were active in founding the National Geographic Society. Hubbard served as its first president from 1888 until his death in 1892. Bell supported Albert Michelson (1852–1931) at Chicago who spent a lifetime measuring the speed of light, and from 1890 Samuel Langley (1834–1906) at the Smithsonian Institution in his research on flight. Bell established a research laboratory in 1883, calling it the Volta Laboratory, and located it first in Washington, DC, and then on his estate on Cape Breton Island in Nova Scotia. In 1907 he organized the Aerial Experiment Association at his Cape Breton estate. It ended by agreement of its members two years later. Bell, who held Unitarian religious beliefs and received 12 honorary doctorates for his contributions to science, died of diabetes.

Electrical industry

Thomas Edison (1847–1931), the key figure in establishing the nineteenth-century American electrical industry, was born in Milan, Ohio, of Canadian parents. His school teacher thought he was addled (retarded), and his mother, insulted by such a characterization, educated her son at home. At age twelve Edison obtained a job selling newspapers on the train running between Detroit and Port Huron, where the family had moved. He became a skilled telegraph operator, the best of whom earned $125 a month, and began a period of wandering from 1863 to 1867 while working as a telegraph operator in different cities. Edison traveled to Boston in 1868 where he invented an electric vote recorder that he hoped the United States Congress would buy but did not, supposedly because its members had no interest in speeding up the voting process as Edison claimed his recorder would do.

In 1869 Edison arrived in New York City, entered into a three-man partnership in Jersey City, and then a two-man partnership called the American Telegraph Company in Newark, New Jersey. In both partnerships he worked on improving stock tickers. In 1870 Edison invented a stock ticker (the printing telegraph) that he sold to Marshall Lefferts (1821–76), an engineer and president of the Gold and Stock Telegraph Company, a subsidiary of Western Union. He supposedly received $30,000 in shares of company stock, although the amount may have totaled $15,000 and Edison's initial share only $1,500. Edison's stock ticker found wide use in receiving stock and commodity quotations in stock exchanges for about 80 years.

Six years later he moved to a 3,000-square foot laboratory building in Menlo Park, New Jersey, where he employed 24–25 workers and where he remained from 1876 to 1887 before moving to West Orange, New Jersey, in the summer of 1887. His West Orange Laboratory consisted of a large 40,000-square foot building and four smaller buildings that occupied 14 acres and employed 100 workers. Edison's laboratories represented the first industrial research laboratories established in the United States. During his lifetime of invention Edison obtained 1,093 patents. The industries based entirely on Edison's inventions, according to the *New York Times*, had a cash value of $15 billion in 1923.

Edison's contribution to the telephone

Edison invented the carbon button variable resistor telephone transmitter in 1877. It was far superior to Bell's diaphragm transmitter, and Edison and Western Union's attempt to use it to circumvent Bell's telephone patent led to the 1877 telephone controversy with Bell. As part of the settlement reached in 1879 the Bell Company paid Edison a $100,000 lump sum payment and $6,000 a year for 17 years, which was the life of the patent. Western Union received 20 percent of the telephone rental income for 17 years. In return Edison and Western Union agreed to leave the telephone business.

Emil Berliner (1851–1929), a German inventor who emigrated from Hanover to Washington, DC, in 1870, invented and filed a patent caveat on 13 April 1877, two weeks before Edison, for a loose contact metal-to-metal, or steel point, pressure-induced variable resistor that he used as a telephone transmitter (Combined Telegraph and Telephone, US Patent 463, 569, filed 4 June 1877, awarded 17 November 1889). Berliner sold Bell the patent rights to his telephone transmitter for $50,000 and worked for Bell as his chief engineer in New York City and Boston from 1877 to 1883 before returning to Washington to become an independent inventor. Berliner in 1897 lost a patent litigation with Edison in a Massachusetts Circuit Court over priority of the variable resistor, although four years later the Supreme Court ruled that Berliner and not Edison had invented the variable resistor telephone transmitter. The Supreme Court also ruled in 1897 that Bell's Berliner patent did not give Bell's company a monopoly on telephone transmitters.

Light bulb

Joseph Swan (1828–1917), an English chemist in Newcastle, began a 30-year period of research on light bulbs in 1848 after learning of an 1845 United States patent that the American inventor John W. Starr (1822–47) in Cincinnati had obtained for a light bulb or lamp. Starr's light bulb consisted of an electrically-heated platinum filament in an evacuated glass container. In other light bulbs Starr substituted a carbon filament for platinum. In his 1848 experiments Swan also tested filaments made of thin strips of carbonized paper, but his bulbs did not perform well. He did not have a steady supply of electricity because only batteries were available, nor could he evacuate the bulb sufficiently to prevent the filament's rapid oxidation and vaporization.

A good-working, practical vacuum pump remained unavailable until Hermann Sprengel (1834–1906), a German who came to Oxford, England, in 1859, built one in 1865. In the older Geissler type vacuum pump that Heinrich Geissler (1814–79), a skilled glassblower in Bonn, Germany, invented in 1855, an open and close tap (faucet) connected the space undergoing evacuation to a mercury barometer (the torricellian vacuum). After partial evacuation by manually raising and lowering a reservoir containing mercury and connected to the barometer by a rubber tube, the operator closed the tap. Repeating the process for as long as two hours produced a good vacuum. The pump worked on the same principle as a force pump operated in reverse. Sprengel mechanized the repeated raising and lowering of the heavy mercury reservoir using a cord and a windlass.

New York Public Library/Photo Researchers/Getty Images

Figure 21 Mercury Vacuum Pump In Edison's Laboratory

Despite these difficulties, Swan experimented with filaments made by carbonizing cotton threads in 1875. He obtained a British patent on his carbon filament lamp in 1880 and began its manufacture in the Swan Electric Light Company that he established in Newcastle in 1881. His lamps lit the House of Commons in 1881 and the British Museum in 1882. Swan's work led to a patent infringement suit with Edison in 1882, which Edison lost, but in 1883, as part of the settlement, Swan merged his smaller company with Edison and established the Edison and Swan United Electric Light Company Limited (Ediswan) in Newcastle, moving to London in 1886.

Although Swan began his light bulb investigations long before Edison, Edison solved the light bulb's main problems, particularly the filament problem on 21 October 1879 (US Patent 223,898, issued 27 January 1880), after trying platinum, gold, and other elements and compounds, all of which melted or vaporized. A carbonized cotton thread sealed in an evacuated bulb worked best as the filament, providing a sufficiently high electrical resistance so that the filament glowed. According to a popular but unsubstantiated legend, Edison's light bulb burned for forty hours continuously, although fifteen hours is closer to the true time.

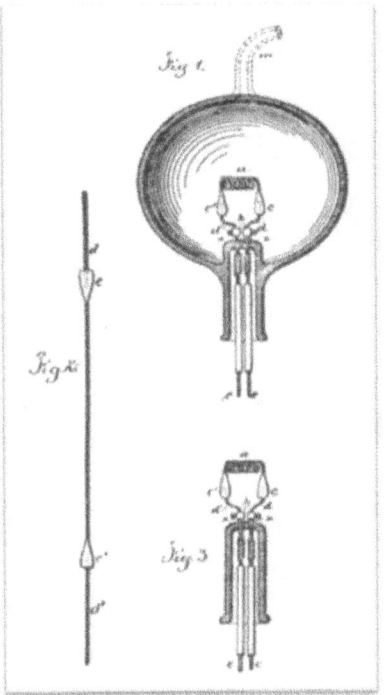

Figure 22 Edison's Electric Lamp

Edison's talented assistants at the time included Francis Upton (1852–1921), a Princeton-educated physicist who joined Edison's laboratory in 1878; John Kruesi (1843–99) a Swiss machinist; James Adams; and Lewis H. Latimer (1848–1928), a self-educated African-American inventor and electrical engineer. Latimer previously worked for Bell making diagrams for Bell's telephone patents and beginning in 1884 carried out research on carbon filaments for Edison. Charles Batchelor (1845–1910) an English textile equipment manager became Edison's right hand man.

Edison's most valued employees such as Batchelor, earned $60 a week, Adams earned $30 a week, and the other workers in Edison's laboratories earned $15–$18 for a 60-hour week plus overtime pay. Skilled workers, such as engineers and machinists, in 1880 averaged $2.17 to $2.45 daily for a 10-hour day. The average annual wage for all workers in 1890 for a 50–60 hour week ranged from lows of $233 for farmers and $256 for public school teachers to $848 for clerical workers in the manufacturing and railroad industries. Federal workers in 1892 averaged $1,096 annually.

Establishing the General Electric Company

Edison owed much of his success to an ability to think and act like a systems engineer. He recognized that for the light bulb or any electrical device to gain widespread public use, he needed an electrical distribution system. Consequently, he developed an electrical grid system that included power switches, meters, lighting fixtures, and fuses to distribute direct current (dc) electricity from his generating station at about 120 volts. In a direct current system the electricity (the electrons) flows only in one direction, from negative to positive electrode, in the electrical circuit. In an alternating current system (ac) the electricity flows in one direction in the circuit, from negative to positive. When it completes its flow through the circuit the electricity reverses its direction of flow and flows through the circuit in the opposite direction. In a 60-cycle ac circuit the electron flow reverses its direction 60 times in one second, in a 120-cycle ac circuit the electron flow reverses itself 120 times in one second. Edison established the Edison Electric Company in New York City, in 1879, and three years later he constructed the first dc commercial electric generating station on Pearl Street in the Wall Street district of New York City.

Omikron/Photo Researchers/Getty Images

Figure 23   Edison DC Generator Long-Legged Mary-Ann 1879

The coal-fed Pearl Street Station began generating electricity on 4 September 1882, distributing electrical current through underground copper wires. His system had 400 light bulbs, each of 50 watts, and 85 customers. Ten pounds of coal burned generated one kilowatt hour of electricity at a cost of 24¢ per kilowatt hour. Edison measured the amount of electricity a customer consumed by using Faraday's laws of electrolysis (1833) to calculate the mass of zinc deposited from one electrode to the other in an electrochemical cell.

CALCULATING THE AMOUNT OF ELECTRICITY CONSUMED

$m = zit$

m = mass of zinc in grams deposited

z = equivalent weight of zinc (atomic weight/2)

i = current in amperes

t = time in seconds

In 1892, ten years after Edison began generating dc current at Pearl Street, his Edison Electric Company merged with the Thomson-Houston Company in Lynn, Massachusetts, to form the General Electric Company in Schenectady, New York. At this time Edison abandoned the use of dc and adopted the more efficient ac generation. The introduction of electrical generating and distribution systems led not only to electric lighting's replacing gas and kerosene lamps, but with the introduction of skyscrapers in the 1870s–80s the electric elevator replaced the steam-powered elevator that Elisha Otis (1806–61) introduced only a few decades earlier in 1857 in New York City. Otis Brothers and Company installed the first electric elevator in 1889 in the Demarest Carriage Building on Fifth Avenue in New York City.

The research tradition that Edison established at the Edison Electric Company continued at the newly established General Electric Company and led to two inventions that produced the modern incandescent light bulb. William D. Coolidge (1873–1975), a chemist at General Electric, developed the tungsten filament in 1911 replacing the carbon filament. Tungsten, which has the highest melting point of any metal (3450° C) and high resistance, gave a bright white light compared to carbon's yellow light and was less brittle.

In 1913 Irving Langmuir (1881–1957), also a General Electric chemist and 1932 Nobel Prize winner in chemistry, introduced the argon gas-filled (or nitrogen-filled) light bulb, thereby eliminating the need to evacuate the bulb. Traces of air (oxygen) in the bulb at high temperatures reacted with (oxidized) the tungsten filament. The pressure of argon, an unreactive gas, retards the vaporization of the filament and increases the light bulb's lifetime and efficiency. Despite these improvements the incandescent light bulb produces a temperature of 2,400° C and still remains only 5–10 percent efficient, the other 90–95 percent energy lost as heat. Its inefficiency has resulted in a global movement to replace the widely-used incandescent bulb with fluorescent lights that are 5–6 times more efficient.

Westinghouse and the introduction of alternating current

An alternative to Edison's dc system emerged in 1886 when George Westinghouse (1846–1914), a Union Civil War veteran, established Westinghouse Electric Company in Pittsburgh and which with his other companies (as many as 34) by 1900 had a capitalization of $120 million and 50,000 workers internationally. Westinghouse Electric developed an ac electrical distribution system with transformers, using a step-up transformer to raise the voltage in electrical transmission and a step-down transformer to lower the voltage in electrical transmission. He borrowed the transformer-electrical transmission idea from his natural gas business where he used gas valves to raise and lower gas pressure. He raised the pressure during transmission from the source (well) and lowered it at the point of use.

The transformer varies the voltage by having a different number of turns of insulated wire wrapped around one side of a rectangular or doughnut shaped soft iron core, called the primary coil, from the number of turns of insulated wire wrapped around a different side of the iron core, called the secondary coil. An alternating current passing through the primary coil produces an alternating electromagnetic field inside the coil. The field passes through the iron core, sweeping back and forth through the secondary coil, and thereby induces an alternating electromotive force or voltage in the secondary coil. The number of turns of wire that the electromagnetic field passes through is what raises or lowers the voltage of the transmitted current. A high number of turns generates a strong field and a high voltage, a low number of turns generates a weak field and a low voltage. The ratio of the two voltages is directly proportional to the ratio of the number of turns.

In commercial electrical transmission, a step-up transformer enables an electric current at low voltage to move from its source through the transmission line at a high voltage of about 7,200 volts to a step-down transformer at the end of the line that lowers the voltage again to about 120 or 240 volts. This method gives the greatest efficiency, about 99 percent in a good transformer, because transmitting electricity at high voltage reduces the current (amperes) flowing in the circuit and thereby the heat energy produced and lost to the surroundings. The quantity of heat produced and lost varies directly as the current squared (Joule's law of 1841). The lower current and heat production enabled ac systems to use copper wire of diameters much smaller than the thick copper wire required for dc transmission.

JOULE'S LAW

$Q = i^2 Rt$
Q = quantity of heat
i = current
R = resistance
t = time

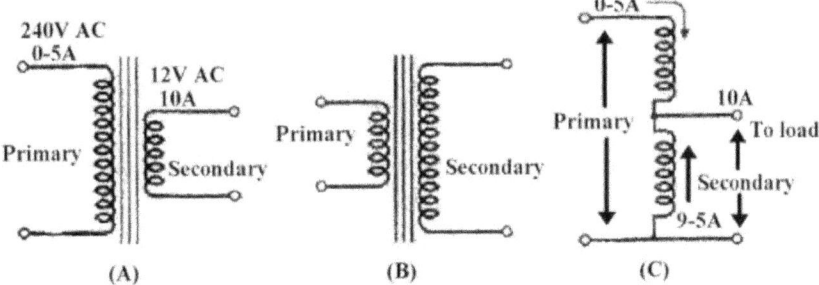

Copyright © EPEmag.Net. Reprinted by permission.

Figure 24 A Simplified Diagram of an AC Distribution System with Step Up and Step Down Transformers (Electromagnetic Induction)

Reduced heat loss, and therefore greater efficiency, is the major advantage of using ac instead of dc because a transformer can vary, raise and lower, the voltage and current only when an alternating electromagnetic field inside the primary coil passes back and forth through the secondary coil. A dc source has a stationary electrostatic field surrounding it. Dc is efficient for transmissions over short distances or about one mile.

Working with Westinghouse on ac generation was the Croatian, Nikola Tesla (1856–1943), the inventor of the ac system who had immigrated to the United States in 1884. For a few years beginning in 1882 Tesla worked for the Continental Edison Company in Budapest and Paris and for Edison in New Jersey but left over a salary issue and a never-received $50,000 monetary compensation for an ac generator design. In 1883–84 he built transformers and in 1888 the first ac generators and first ac polyphase system in which three separate currents or waves of electricity having the same frequency are out of phase. Consequently, the three conductors carrying the alternating currents reach their peak values at different times. The phase delays of 1/3 of a cycle in one conductor and 2/3 in the second conductor, and of no delay in the third gives constant power transfer over each cycle of the current and made possible an induction motor in which the magnetic field rotated rather than the motor's armature electromagnet. The rotating magnetic field eliminated the sparking of the older metallic commutator-carbon brush ac motor. The three-phase system powered large induction motors and other large loads. It had a higher efficiency than single- or two-phase systems, transmitting the same power load with fewer conductors. For use of these inventions Westinghouse in 1888 paid Tesla only $60,000 in cash and stock and $2.20 for each horsepower his system generated.

Westinghouse and his chief engineer in Pittsburgh, the physicist William Stanley (1858–1915) inventor of the induction coil for high voltage production (US Patent 349,681 awarded 21 September 1886), constructed the first commercial ac system in the village of Great Barrington, Massachusetts, on 30 November 1886. In 1893 he convincingly demonstrated ac's advantage over dc at the World's Columbian Exposition in Chicago. Mainly because ac's electrical distribution efficiency was higher than dc's, Westinghouse, in competition with Edison, also won the Niagara Falls Power Company's $100,000 prize for the development of an electrical generation system at Niagara Falls, New York, then a village of 5,000, located on the Niagara River, which separates Canada and the United States.

Westinghouse and Tesla began construction on an ac generating station with ten 5,000 horsepower generators, 430 cubic feet of water diverted from the Niagara River turned the turbines at 250 revolutions per minute. They completed the generating station in August 1895 and in November 1896 extended the electrical transmission to Buffalo, New York, a growing industrial center of 250,000, 22 miles away. The panic of 1897 led to Westinghouse's and his companies' large financial losses and to losing control of his companies in 1911. He died of heart failure in 1914.

The decade-long ac-dc competition with Westinghouse and Tesla beginning in 1886 revealed Edison at his worst. In an 1887 pamphlet *A Warning* Edison tried to frighten the public by exaggerating the danger of ac. The next year he resorted to scare-tactic electrical exhibitions in the Dynamo Room of his West Orange laboratory. The electrical exhibitions were demonstrations from July to December 1888 in which Edison's associate, the engineer Harold P. Brown (1869–1932), used a 1,000–1,200 volt ac generator to electrocute small and large animals, including 50 dogs and cats, a 125-pound and 145-pound calf, and a 1,230-pound horse. Brown gave the 125-pound calf, which had a body resistance of 3,200 ohms at the place of application, 700 volts for 30 seconds, and killed it instantly. The 145-pound calf had a resistance of 1,300 ohms, and 700 volts for 5 seconds killed it instantly. The horse, which had a resistance of 11,000 ohms, received 700 volts and lasted 30 seconds.

The electrocutions, intended to demonstrate dramatically how easily ac killed, were the outcome of a fatal accident seven years earlier. On 7 August 1881 Alfred Southwick (1826–98), a dentist and onetime steamboat operator in Buffalo, saw an inebriated old man fall on or touch the terminals of a dc electrical generator at the Brush Electric Light Company where Brown once worked, and experience an apparently quick and painless death. Southwick, who had studied electricity and thought of testing it as a possible anesthetic, believed that the speed with which electricity killed provided a more humane method of eliminating unwanted stray animals. Beginning in 1882 he and a colleague, the Buffalo physician, George Fell (1850–1918) who also taught medicine at nearby Niagara University, began a long series of experiments to test electrocution's effectiveness.

Southwick concluded from his animal tests that for criminal convictions warranting the death penalty, electricity provided a quicker and more humane execution than hanging. Hanging was by far the most common method of execution in the 1880s, but because it often resulted in slow gruesome strangulations or decapitations, New York State and several other states were looking for alternative ways to apply the death sentence.

Southwick's friend, Daniel McMillan (1846–1908), a New York state senator, although a strong advocate of capital punishment, acknowledged the arguments of those who opposed capital punishment as cruel and inhumane. When Southwick showed MacMillan his experimental results, especially the quick and probably painless animals' deaths, MacMillan believed that electrocution would end the ongoing capital punishment debate. He convinced New York State's Democratic governor David B. Hill (1843–1910) that Southwick's results were reliable, and in an

1885 address Hill directed the state legislature to find a less barbaric method of criminal execution.

The legislature failed to act, and in 1886 Hill appointed a three-member *Commission to Investigate and Report the Most Humane and Practical Method of Carrying into Effect the Sentence of Death in Capital Cases*. Its members, Southwick, Matthew Hale an Albany judge, and Elbridge Gerry (1837–1927) legal advisor to the American Society for the Prevention of Cruelty to Animals, examined the few alternatives to hanging, such as a firing squad, and in January 1888 they submitted a 95-page report recommending that electrocution replace hanging as the state's method of execution. The state legislature on 4 June 1888 passed Chapter 489, the law establishing electrocution as the method of execution, and designated the state's Medico-Legal Society to determine how to implement the law. On 6 June 1888 Hill signed New York State's Electrical Execution Law, the world's first electrocution law.

The undecided issue, and the reason for the Edison-Westinghouse controversy, was whether to use ac or dc in electrocution. Southwick had corresponded with Edison in 1887 while conducting his electrical experiments, and in their correspondence Edison promoted Westinghouse's ac generators as more effective than dc generators. Brown's ac electrocutions in Edison's laboratory convinced Brown's friend Frederick Peterson (1859–1938), a medical doctor at Columbia and head of the Medico-Legal Society, and other society members who witnessed them in November 1888. The Society's report, submitted in December, recommended that the state adopt ac for electrocutions. The report also recommended that the condemned sit in an electric chair rather than on a rubberized table or in a water tank during the electrocutions.

New York State's execution law went into effect on 1 January 1889. To implement the ac electrocution system General Austin Lathrop (b. 1839), superintendent of New York State's prison system from 1887 to 1898, tried to purchase three Westinghouse ac generators in March 1889. But because Westinghouse refused to sell his generators directly to the prison system, Brown and Edison schemed to acquire three generators by having them shipped to Brazil and then back to the United States supposedly for $7,000–$8,000.

Joseph Chapleau (1850–1910), a poor French-Canadian farmer in Plattsburgh, New York, holds the title of first person sentenced to death by electrocution, but escaped execution. Chapleau brutally killed Erwin Tabor, a wealthy neighboring farmer, in

1889 by beating him on the head with a four-foot sled stake because he believed Tabor had poisoned two of Chapleau's cows with rusty nails and degraded his own (Tabor's) female servants by demanding sexual favors from them. Fortunately for Chapleau the court commuted his death sentence to life imprisonment in Dannemora Prison in the upstate New York village of Dannemora.

The dishonorable title of first victim fell to Buffalo's William Kemmler, alias John Hart (1862–90), an illiterate, alcoholic, vegetable peddler, for the March 1889 hatchet murder of his girlfriend, or common law wife, Matilda (Tillie) Ziegler, whom he struck 26 times in the head. Kemmler believed she planned to leave him for another man.

Following Kemmler's conviction and death sentence Westinghouse paid for Kemmler's unsuccessful appeals to New York State's courts and the United States Supreme Court. Westinghouse believed electrocution was a cruel and unusual punishment and did not want his name associated with or used synonymously with electrocution. The appeals failed, and in 1890 Edwin Davis (1846–1923), the first state electrician at Auburn Prison, designed the first electric chair, constructing it of oak with leather straps and testing its efficiency on large slabs of meat.

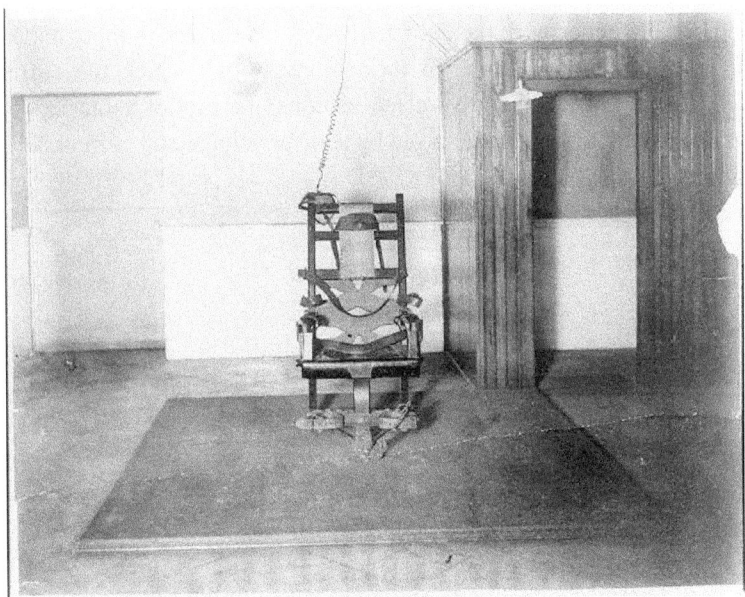

Courtesy of the Library of Congress

Figure 25   The Electric Chair in Auburn State Prison

Despite Davis's tests Kemmler's electrocution before 25 witnesses in the prison's basement on 6 August 1890 was a disaster. The first application of ac, 1,400 volts and a probable alternating current of 3 amperes for 17 seconds, failed to kill Kemmler, who had electrodes attached to his head and spine. He received another 1,400 volts for two and a half minutes before succumbing, being literally roasted to death.

Brown, assisted by Peterson, and later Fell, designed electric chairs for three New York State prisons, at Auburn, Sing Sing, and Clinton. From 1890 to 1916, 55 electrocutions occurred at Auburn. In 1916 all state electrocutions took place at Sing Sing. From 1890 to 1963 when New York State abandoned the electric chair, 686 men and nine women died in the chair. Other states soon introduced electrocution: Ohio in 1896, Massachusetts in 1898, New Jersey in 1906, Virginia in 1908, and North Carolina in 1910. Davis between 1890 and 1914 reputedly electrocuted 240 prisoners in New York State, New Jersey, and Connecticut, among them Leon Csolgosz (1873–1901), President William McKinley's assassin.

Because the first electrocutions were not always swift and painless those states that adopted electrocution almost immediately after Kemmler's electrocution increased the voltage to 2,000 volts and an alternating current of five amperes. The increase proved effective because the human body's resistance, although not constant throughout, is 400–500 ohms in the region through which the current passes. Usually the victim received two electrocutions, the first of about 2,000 volts and 5–7 amperes for 30 seconds followed by a lower voltage and current, 240 volts and 1.5 amperes for 60 seconds.

OHM'S LAW
$V = iR$
V = voltage (volts)
i = current
R = resistance

Edison, who said that he opposed capital punishment, promoted ac electrocution in order to horrify and alert the public to the danger of ac and turn the public against ac in favor of dc. He suggested calling electrocution *Westinghoused*. *Electromort*, *dynamort*, and *electricide* were other suggested names for electrocution. Edison's plan backfired because the Niagara Falls Power Company adopted ac for its commercial electrical generation, and New York State did the same for electrocutions. Critics of electrocution attacked the new method of execution as cruel and unusual punishment and in violation of the United States constitution, but both state courts and federal courts upheld the state law.

Edison's other inventions

Phonograph

The idea of the phonograph, an instrument that could record a message and which Edison said was his most original invention, came to him in 1877 while he was thinking of a way to store the words spoken into a telephone. Bell and Western Union, Edison's financial backer in the telephone controversy with Bell, at first doubted the telephone's economic impact. Western Union's business was receiving and sending telegraph messages and delivering telegrams, not voices. Edison, nevertheless, thought the phonograph might provide another way of delivering a message, that of a recorded human voice, and thereby interest Western Union.

Edison had worked with thin metal diaphragms while developing the telephone's carbon granule microphone. He knew that a good diaphragm vibrated at the same frequency as the sound waves directed against it and reproduced the human voices that went into and came out of the telephone. While working on the design of an instrument that recorded and stored the transmitted sounds, Edison attached a straight pin, or needle, perpendicular to one side of the diaphragm so that when he directed his voice against the other side of the diaphragm the pin moved up and down, vibrating at the same frequency as the sound waves it received.

In the first test of his recording instrument, Edison placed a ribbon of waxed paper beneath the pin. He slowly pulled the paper so that the pin made an indentation lengthwise in the wax, and at the same time he said *hello*. The resulting indentation was the wax recording. Edison then placed the wax recording beneath a second wax-free pin, set the pin at the beginning of the waxed indentation, and pulled the paper under the pin so that the pin followed the path of the indentation. Upon doing so, he and Batchelor heard what sounded like *hello*. The recording instrument worked.

The first phonograph that Edison built in December 1877 consisted of tin foil wrapped around a 3½ inch rotatable horizontally-arranged solid brass cylinder. A metal needle (stylus) attached vertically to a diaphragm fitted tightly over the smaller opening of a megaphone placed above the cylinder. The needle made grooves or impressions in the foil when sound waves from a voice or other source passed through the megaphone and caused the diaphragm and the needle to vibrate. Upon rotating the cylinder, the up-and-down vibrating needle, by means of a worm gear, passed down the foil-covered cylinder making grooves that represented the frequencies of the vibrating needle. To play back the sound, the operator returned the needle to the beginning of the grooves and rotated the cylinder. This caused the needle and the diaphragm to vibrate at the frequencies impressed in the grooves, thereby playing back the original sound. Edison's first phonographs were not electrical but hand-operated and cost about $15 each.

Philip G. Hubert (1896). *Inventors.* New York, Charles Scribner's Sons

Figure 26   Edison's First Phonograph

Mary had a Little Lamb was the first song Edison recorded. He and his assistants chose a nursery rhyme because they had young children. When Edison first demonstrated his phonograph in public, skeptics said it was a trick ventriloquist and not a recording machine. Edison prevailed and soon recorded himself, the German statesman Otto Bismarck (1815–98), the British Prime Minister William E. Gladstone (1809–98), and many others. On 23 November 1889 the world's first coin-operated (5¢) Edison phonograph appeared in the Palais Royale saloon in San Francisco. By the 1890s phonograph parlors were popular throughout the country.

Copyright © Science Museum / Science & Society Picture Library

Figure 27   Edison Records Himself

Emil Berliner in 1887 made the most significant improvement to the phonograph when he invented the flat phonograph shellac disk or record. In Berliner's record design the needle vibrated from side to side in grooves as the record turned at 78 revolutions per minute and not up and down as in Edison's tin foil-covered cylinder. Berliner in 1890 founded the Berliner Gram-O-Phone Company in Waltershausen, Germany, because most Americans preferred Edison's cylinder to his flat record. He established company branches in Washington, DC, in 1892 which in 1895 had a $25 million capitalization, in London in 1898, and in Montreal, Canada, in 1899. Berliner's flat record's greater compactness and practicality enabled it to replace Edison's cylinder in the early 1900s.

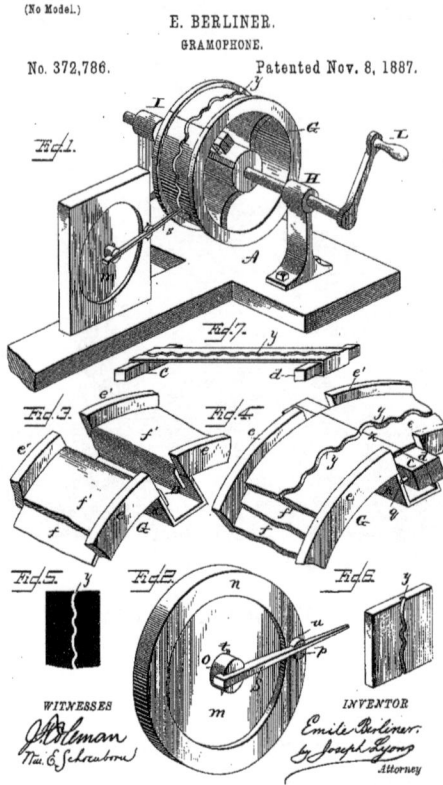

Figure 28 Berliner's Gramophone

Berliner's company manager in London, William Barry Owen (1860–1914), introduced the Berliner Company's famous trademark in 1900. In 1899 he paid the British artist Francis Barraud (1856–1924) £50 to modify Barraud's 1898 painting of Barraud's fox terrier Nipper (1884–95), *Dog Looking at and Listening to a Phonograph*, now known as *His Master's Voice,* and £50 for the copyright ($500 total). Nipper remained Berliner's trademark after his company merged in 1901 with the Consolidated Talking Machine Company of Eldridge Johnson (1867–1945) to form the Victor Talking Machine Company in Camden, New Jersey. Radio Corporation of America (RCA), established in 1919 in Riverhead, Long Island, bought Victor in 1929 for $145 million. Victor and RCA maintained the Nipper trademark.

The Columbia Phonograph Company, founded in 1888 in Washington, DC, in 1901 made the first commercial flat records and cylinders featuring artists such as the American composer John Phillips Sousa (1854–1932) and his marching band. It originally distributed and sold Edison's cylinders, but beginning in 1912 it sold only flat records. The Gramophone Company (Victor Talking Machine Company) in London recorded the first 10 flat records of the Italian opera star Enrico Caruso (1873–1921) in Milan on 11 April 1902. Caruso's fee was $10,000 in 2000's dollars.

Motion picture machine

Edison constructed a motion picture machine in 1889. He took a series of photographs on a strip of Eastman film and then flashed the photographs on a screen in rapid succession. Perforations along the sides of the film, through which sprocket wheels turned, moved the film at a uniform speed past a bright source of light. Edison made several early films of boxing matches in his *Black Maria* studio built in West Orange in 1892–93, including an 1894 bout between Peter Courtney and Gentleman Jim Corbett (1866-1933), heavyweight champion 1892-97; and the *Gordon Sisters Boxing* (Bessie and Minnie) in 1901. At Coney Island, Brooklyn, New York, Edison filmed without authorization the 1899 heavyweight title bout in which the champion James Jeffries (1875-1953), heavyweight champion from 1899 to 1905, defeated by decision Tom Sharkey (1873-1953). He also made movies such as *The Great Train Robbery* (1903), *Uncle Tom's Cabin* (1903), and *The Prince and the Pauper* (1909). They were full-length movies, which at that time meant about 14 minutes. Edison soon lost interest in movie making and left its development to others.

Spirit communication machine

Edison discussed constructing a spirit communication machine shortly after World War I, at a time when the spiritualist movement was re-emerging in England and Europe. The tremendous loss of life, military and civilian, estimated as high as 20 million people, many of them quite young, drove the movement and influenced several scientists, among them the physicist Oliver Lodge (1851–1940) at Birmingham, the physician Arthur Conan Doyle (1859–1930) in London, and the physicist William Barrett (1844–1925) in London. The latest movement followed an 1870s–80s spiritualist movement that included the philosopher Henry Sidgwick (1838–1900) at Cambridge, the naturalist Alfred Russel Wallace (1823–1913) in London, the physicists Lord Kelvin (1824–1907) at Glasgow and Pierre Curie (1859–1906), Marie Curie's husband, in Paris, and the chemists Karl Kellner (1851–1905), a freemason in Vienna, and William Crookes (1832–1919) in London.

Edison, a Deist whose God was nature and who opposed all organized religion, earlier in 1878 had come under the influence of the Russian occultist Madam Helena Petrovna Blavatsky (1831–91), who blended spiritualism, Buddhism, and Hinduism and in 1875 founded the Aryan Theosophic Society. She tried to assimilate science with theology. Edison met her when she resided in New York City from 1873 to 1878. He discussed a spirit communicating machine in interviews in October 1920 in *The American Magazine* and *Scientific American*. The machine consisted of a highly sensitive valve that detected a spirit's weakest whisper, utterance, movement, or effort; magnified or amplified the effect as in a megaphone; and then recorded it. But he and Miller Hutchinson (1876–1944), his chief engineer at the West Orange laboratory and employee from 1912 to 1917, never built such a machine. Edison's seriousness remains doubtful because of the many exaggerations and distortions about his life. His biographers do not mention an Edison spirit communicating machine.

CHAPTER 2: The Technological Transformation   81

\multicolumn{4}{c}{**TECHNOLOGICAL BREAKTHROUGHS 1850s–1920s**}
Industry	*Technological Breakthrough*	*Year of Invention or Patent*	*Contributor*
Steel	Pneumatic furnace (air blast)	1851 (I)	William Kelly
	Bessemer converter	1856 (P)	Henry Bessemer
	Spiegeleisen (iron alloy)	1856 (P) US 1857 (P)	Robert Mushet
	Basic furnace lining	1877–78 (P)	Sidney Thomas and Percy Gilchrist
	Open hearth process	1856–61 (I)	Friedrich and William Siemens
Aluminum	Electrolytic process	1889 (P)	Charles Martin Hall
	Bauxite (alumina) in cryolite	1889 (P)	Louis Héroult
Petroleum	Fractional distillation	1855 (I)	Benjamin Silliman, Jr.
	Pipe-driving method	1859 (I)	Edwin Drake
	High-pressure thermal cracking	1913 (I)	William Burton
	Catalytic cracking	1915 (P)	Almer McAfee
	Catalytic cracking	1927 (I)	Eugene Houdry
Communications	Electromagnetic induction	1876	Alexander Graham Bell
Telephone	Conservation of energy		
	Carbon granular resistor	1877 (P)	Thomas Edison
	Toroid induction coil	1894 (I)	Michael Pupin
Phonograph	Diaphragm vibration	1877 (I)	Thomas Edison

TECHNOLOGICAL BREAKTHROUGHS 1850s–1920s			
Industry	Technological Breakthrough	Year of Invention or Patent	Contributor
Electrical	Carbon filament-evacuated light bulb	1879 (I)	Thomas Edison
	AC distribution system	1886 (I)	Nikola Tesla and George Westinghouse
	Transformer, ac motor, generator	1883–88 (I)	Nikola Tesla
	Tungsten filament	1911 (I)	William Coolidge
	Argon-filled bulb	1913 (I)	Irving Langmuir

A brief look at the emerging American chemical industry

In addition to the chemical advances made in the iron and steel, aluminum, and petroleum industries, important chemical developments occurred in the salt, borax, sulfur, electrolytic, and plastics industries.

Salt industry

The 1849 California gold rush and the westward pioneer movement greatly increased demand for salt (NaCl), the only safe food preservative then available. This resulted in the flourishing of Richmond and Company, established in 1848 as a salt sales company in Chicago, and of Joy Morton (1855–1934) who in 1886 acquired a majority interest in the company and renamed it Joy Morton and Company. The company began to acquire salt plants in 1890 and in 1910 became incorporated as the Morton Salt Company.

To extract salt from underground deposits miners drilled two parallel wells to the deposit and connected them by a horizontal tunnel. Pumping water down one well dissolved the salt and drove the salt solution up the second well to the surface, leaving the insoluble impurities behind. By the 1830s the salt (and the sugar) industry had adopted an evaporation technique in which the salt solution passed from the well into large pans. Boiling off the water left behind the purified salt.

Later improvements in salt production included the addition of magnesium carbonate in 1911 to prevent caking. The addition of potassium iodide in 1924 prevented goiter, a disease of the thyroid gland, resulting in its swelling, and visible in the front of the neck. Morton was among the first salt companies in the United States to make these improvements. Its famous slogan "when it rains it pours" first appeared in *Good Housekeeping* in 1924.

Morton established the Morton Arboretum on 14 December 1922. He transformed his 735-acre estate (now 1,700 acres), 25 miles west of Chicago, into a center for horticulture, shrub, plant, and tree studies and research.

Borax industry

John A. Veatch (1808–70), a medical doctor and chemist, in 1856 discovered white borax crystals ($Na_2B_4O_7.10H_2O$) at Borax Lake in Lake County, north of San Francisco. He made a second discovery in 1860 at Little Borax Lake four miles west of Borax Lake. Veatch bought the land from its Indian owners because of borax's known use as a cleanser and preservative, and for its healing properties.

The California Borax Company, which five investors established at Borax Lake in 1864, produced 12 tons of borax in its first year of operation and 590 tons between 1864 and 1868. The company moved its operation to Little Borax Lake in 1868, where from 1868 to 1873 it provided the United States with its entire borax requirement, 140 tons valued at $89,100. The discovery of large borax deposits in Nevada in 1870 and in California's Death Valley in 1881 ended all borax production in Lake County. The five owners of the California Borax Company changed the company's name in December 1875 to the Sulphur Bank Quicksilver Mining Company and continued operation as quicksilver (mercury) and sulfur producers.

Nevada's large-scale borax production began in 1870 with the discovery of *ulexite*, a white crystalline sodium calcium hydrated borate, $NaCa[B_5O_6](OH)_6.5H_2O$, known as cottonball, in Columbus Marsh in western Nevada. A few years later in 1872, Francis Marion "Borax" Smith (1846–1931) established the Pacific Coast Borax Company to develop this site and nearby Teel's Marsh. William Tell Coleman (1824–93), a vigilante enforcer who acquired fame in the 1850s for making San Francisco a safe city, was Smith's business associate and distributor until Coleman's bankruptcy in 1888.

Coleman entered the borax business when he bought for $20,000 a large cottonball deposit that Aaron and Rose Winters in 1881 had discovered in the Furnace Creek area of California's Death Valley. The next year he acquired two more borax deposits. Coleman Company prospectors in Death Valley found deposits of a richer white borax mineral, calcium borate, that they named colemanite, $CaB_3O_4(OH)_3 \cdot H_2O$, and he bought for $4,000 a large colemanite deposit that three independent prospectors discovered a few miles away at Mont Blanc.

As a result of these discoveries, and subsequent discoveries, Coleman in 1882 constructed the Death Valley Harmony Borax Works 1.5 miles north of the mouth of Furnace Creek to mine the borax deposits. Its laborers, mainly Chinese, worked a 12-hour day, 7-day week, and in 1883 earned $45 for shoveling the smelly raw borate ore from the surface of the ground into massive 10-ton capacity wagons that cost $900 each. Two ten-mule teams pulled the wagons across 165 barren miles of California desert to the nearest railroad junction at Mojave. The round trip took 20 days. This is the origin of the famous 20-mule team borax symbol that Coleman designed and used in 1891 but which remained unregistered until 1894. From 1883 to 1888 when the Harmony Borax Works went bankrupt, the 20-mule team hauled 20 million pounds of borax out of Death Valley. Railroads later replaced the mule teams and in 1898 were transporting 150 tons a day.

Borax found use at the time in soaps and hand cleansers, as a water softener, a complexion aid, a digestive aid, and in treating epilepsy and bunions. With the collapse of Coleman's borax business, Smith, to whom Coleman was heavily mortgaged, acquired Coleman's Harmony Borax Works and his Amargosa and Almeda plants and refineries. Smith consolidated these in 1890 as the Pacific Coast Borax Company which in 1956 became United States Borax and Chemical Company.

Sulfur industry

Herman Frasch (1851–1914), a chemical engineer who in 1868 emigrated from Halle, Germany, to Philadelphia, in 1885 invented a process for desulfurizing crude oil and reviving exhausted oil wells by treatment with hydrochloric acid. Beginning in 1891, while working as chief chemist for Standard Oil in Cleveland, Frasch developed an extraction process for sulfur mining that made available Louisiana's significant, but quicksand-covered, sulfur deposits and lessened Sicily's (Italy) control of sulfur production. Frasch's extraction process used three concentric pipes driven into the sulfur deposits. He sent superheated water down the outer pipe to melt the sulfur and forced compressed air down the central pipe. This pushed the molten sulfur up the middle pipe for collection in storage containers. The Frasch process increased sulfur production from 7,000 tons in 1901 to 220,000 tons in 1905. It still produces much of the global sulfur supply.

Electrolytic industry: Dow Chemical Company

Herbert H. Dow (1866–1930), a Canadian (Ontario) by birth, graduated in 1888 from the Case School of Applied Science in Cleveland where he studied with Edward Morley (1838–1923) and Albert Michelson (1852–1931). In 1889 he patented an electrolytic process that removed bromine from underground brine (salt water) solutions. His process converted the metallic bromides in brine to liquid bromine for conversion into bromine compounds, such as sodium bromide and potassium bromide, first used in photography and medicine and later in flame retardants and water purification. It was the second electrolytic process in the American chemical industry to use direct current generators. Hall began electrolysis of alumina in November 1888.

Dow established the Midland Chemical Company in 1890 and in 1895 a second company, the Dow Process Company in Canton, Ohio, to electrolyze brine for chlorine. He moved his Ohio company to Midland in 1897 where he merged the two companies and established the Dow Chemical Company. In 1905–06 Dow broke the German chemical cartel Die Deutsche Bromoconvention's monopoly on bromine production and world price. Beginning in 1916 company products expanded to include synthetic indigo (1916), calcium and magnesium metal, Epsom salt (magnesium sulfate), iodine, and mustard gas during World War I. A prolific inventor and innovator, Dow obtained over 90 patents.

Plastics industry

The chemist Leo Baekeland (1863–1944) came from Belgium to New York City in 1889, after spending time in France and Britain. Earlier in his career, in 1893, Baekeland invented a photographic print paper, *velox*, that enabled photographers to develop their prints in artificial light. George Eastman (1854–1932) in 1899 bought the rights to *velox* for about $750,000. Twenty years later, in 1909, Baekeland invented *bakelite*, the first durable plastic, by reacting phenol with formaldehyde, and in 1910 he established the General Bakelite Corporation in Perth Amboy, New Jersey, to manufacture bakelite. Baekeland sold his company to Union Carbide and Carbon Company in 1939, the year World War II broke out in Europe. Bakelite and later plastics accomplished what Goodyear tried to do with rubber. The plastic boom began after World War II and continues, with no apparent limit to the use of plastics.

Automobile industry

The first self-propelled vehicles or automobiles were the steam-powered vehicles developed in Europe and the United States in the late 1770s and early 1800s. But because of the steam engine's low efficiency, about 17 percent, and inconvenience the gasoline-powered internal combustion engine replaced the external combustion steam engine. Europe gave birth to the internal combustion engine particularly through the contributions of the German inventors Karl Benz, Nikolaus Otto, and Göttlieb Daimler, whereas the automobile industry, or mass production of automobiles, began in the United States in the early twentieth century. Ransom Olds and Henry Ford applied the assembly line method of firearms and clock production to automobile production.

External combustion engine: steam-powered and internal combustion gasoline and diesel-powered vehicles

Nicolas Cugnot (1728–1804), a military engineer in France, in 1770 invented the first steam-powered vehicle. His three-wheeled, two-cylinder vehicle carried four passengers and had a speed of two to three miles per hour. Oliver Evans (1755–1819) in Philadelphia in 1804 built the first steam-powered vehicle in the United States.

The first internal combustion engine appeared 50 years later. Etienne Lenoir (1822–1900), a self-educated engineer in Luxembourg and France, in 1854 built an internal combustion engine that ran on illuminating gas, a mixture of hydrogen and methane. The fuel passed into one end of the engine's single cylinder where it mixed and burned with air admitted at the cylinder's other end. His engine operated on the same principle as Watt's reciprocating steam engine having cycles of induction, explosion, and expansion, alternating with exhaustion on either side of the piston. Small industrialists found Lenoir's engine, which developed about 0.5–2.5 horsepower, very useful, and in the five years after Lenoir obtained a patent (1860–65) they bought about 300 of his engines. Lenoir attached his engine to a carriage to produce locomotion in 1860. In 1883 he built a four-stroke engine.

In Germany, Karl Benz (1841–1929) in Mannheim, after experimenting with an internal combustion (gasoline) engine for most of the 1870s and receiving a US patent in 1877, significantly improved the internal combustion engine by 1879. In

1886 he designed and patented a 0.8 horsepower engine for automobile use. Benz continued to improve his engine, and by 1899 he built 2,000 four-wheel automobiles.

Seven years before Lenoir built his four-stroke piston engine Nikolaus Otto (1832–91) at Gasmotoren-Fabrik in Deutz-Cologne, built a four-stroke piston sequence engine that had a higher compression ratio and ran much more quietly than Lenoir's engine. Otto had his 1877 German and US patent invalidated in 1886 when his French competitors claimed patent infringement and prior invention. They noted the engine design the French engineer Alphonse Eugene Beau de Rochas (1815–93) in Digne-les-Bains described in a pamphlet he published in 1861 and patented in 1862 and the designs of Christian Reithmann (1818–1909) a Munich watchmaker who in 1860 received a one-year German patent. Beau de Rochas never built a four-stroke engine nor pay his patent tax, Reithmann's patent claims in 1860, and designs in 1867 and particularly in 1873, have remained unconvincing and controversial. Nevertheless, within ten years Otto sold about 30,000 of his more efficient and quieter Otto cycle engines.

Gottlieb Daimler (1834–1900) and his partner Wilhelm Maybach (1846–1929) worked with Otto at Deutz. They left in 1882 and that same year established a company in Stuttgart to manufacture light, high-speed efficient internal combustion (gasoline) engines. They attached one of their engines to a bicycle in 1885, giving the first motorcycle, and a two-cylinder engine to a four-wheeled vehicle in 1887, giving the first practical automobile. Daimler and Maybach founded the Daimler Motor Company in Stuttgart in 1890 but split over policy differences. Maybach left in 1907 and established his own company in Friedrichshafen.

An alternative to the gasoline piston engine appeared in 1892 when Rudolf Diesel (1858–1913), a French engineer in Germany, patented the diesel (compression) engine. Diesel at various times had plants in Augsburg, in St. Petersburg with the Nobel family, and in France. The engine he designed ignited its fuel (diesel oil) by compressing it to produce ignition, rather than sparking as in a gasoline internal combustion engine. Diesel engines were 50–56 percent efficient operating between 1527° C and 527° C, which was more than the 39 percent efficiency of a gasoline internal combustion engine operating between 1000° C and 500° C. The diesel engine in the twentieth-century became standard on railroads, trucks, and buses.

American automobile manufacturers and organizers

Charles Duryea (1861–1938) and his brother Frank Duryea (1869–1967) in Springfield, Massachusetts, built the first successful American automobile, demonstrating it in September 1893. Their automobiles had interchangeable parts, a water-cooled, two-cylinder, four-cycle, 3.5 horsepower engine, and achieved a top speed of 7.5 miles per hour. The company they founded in 1895, and in 1896 sold the first American automobiles, selling 13 that year, lasted only three years before the brothers parted company and built automobiles separately.

Another American auto pioneer, Ransom Olds (1864–1950) in Detroit-Lansing, obtained his first patent on a gasoline-powered vehicle in 1886. He established the Olds Motor Vehicle Company in 1889 and began assembly line production of the *Oldsmobile Curved Dash Runabout* in 1901, producing 425 Oldsmobiles. Production increased to 6,500 Oldsmobiles annually by 1905. By 1903 Olds had agents selling cars in 18 countries and was the first U.S. automobile company to export automobiles.

Olds' first automobiles, built from 1901 to 1903, weighed 650–700 pounds, had a 4-horsepower, 1-cylinder engine, and reached a top speed of 25–28 miles per hour. Engine power increased to seven horsepower in 1904, but the Oldsmobile's $650 cost remained the same. Oldsmobile had 37 percent of the market in 1903 but fell to 5 percent in 1906. Olds left the company because of a dispute with his backers and established the Reo Motor Company in 1904, serving as its president from 1904 to 1924. Oldsmobile became a division of General Motors Company in Flint, Michigan, in 1908.

The Ford Motor Company, established in 1903, was a major player in the automobile industry. Henry Ford (1863–1941) set up an assembly line in 1909 in Dearborn, Michigan. It decreased production time from 14 hours to 93 minutes and reduced the price of a Model-T from $950 in 1909 to $600 in 1912 when the average annual earning for all industries was $592–650. Ford introduced an 8-hour, $5-per day work week in 1914. He doubled the prevailing average annual wage of $650, a strategy that enabled his workers to purchase a Ford automobile and reduced worker turnover. Other manufacturers would have to raise their workers' wages to remain competitive. In 1912 Ford sold 40,000 cars and had 20 percent of total sales. In 1921 it sold 945,000 cars for a 55 percent share.

The United States' largest automobile manufacturer, General Motors Corporation (GMC) that William C. Durant (1861–1947) first organized in 1908 in Flint, and then called General Motors Company, underwent reorganization after Durant's forced departure in 1920. Alfred P. Sloan (1875–1966), who joined GMC in 1908, became president and Charles Kettering (1875–1958), head of GMC Research Corporation, its vice-president. Both were engineers.

Kettering founded Dayton Engineering Laboratories Company (Delco) in Ohio in 1909. GMC bought the company in 1918. He held many automobile-related patents, including the first electrical ignition system and self-starter for automobiles and the first practical engine-driven generator both of which he introduced in the 1912 Cadillac. Kettering also worked with Thomas Midgley (1889–1944) on the discovery in 1921 of tetraethyllead, a gasoline additive that prevented engine knock, and in 1930 on the preparation of freons (chlorofluorocarbons), a coolant in refrigerators and air conditioners. Neither remains in use today. Tetraethyllead puts dangerous amounts of the toxic element lead into the atmosphere and freon attacks the atmosphere's ozone layer allowing greater amounts of ultraviolet radiation to reach Earth.

A polio attack in 1940 paralyzed Midgley. He designed a special type of harness with pulleys to give him some maneuverability, but the harness led to his strangulation and accidental death in 1944. Sloan and Kettering established the Sloan-Kettering Institute for Cancer Research in New York City in 1948.

Walter Chrysler (1875–1940) was plant manager at the American Locomotive Company in Schenectady, New York, and in 1919 president of the Buick Motor Company. He built the first Chrysler automobile in 1924 and in 1925 transformed the dying Maxwell Motor Company (1903–25) and Willys-Overland Company into the Chrysler Corporation in Detroit. GMC, Ford, and Chrysler were the *big three* of the American automobile industry. By 1930 they controlled 83 percent of the market.

CHAPTER 3

INDUSTRIAL ORGANIZING METHODS OR TECHNIQUES AND INDUSTRIAL ORGANIZERS

I. GENERAL CHARACTERISTICS OF MAJOR INDUSTRIES

Size, wealth, and power were the most striking characteristics of the new American industrial giants. During the last half of the nineteenth century, limited partnerships and individual or family-owned businesses, such as Bell's, Edison's, and Hall's, gave way to large corporations owned by stockholders. These corporations possessed considerable wealth and had the capital required to expand and to acquire the industrial-scale apparatus and equipment needed to remain competitive. Their size and wealth enabled them to have significant influence both economically and politically.

Expansion, usually followed by consolidation and monopoly or oligopoly, characterized the nineteenth-century American technological transformation. As a result of the tremendous industrial expansion several organizational methods, such as patent pools, trusts, and cartels created for the sole purpose of gaining partial or complete control of a particular industry, emerged and dominated American industry for a time. Integration, either vertical or horizontal, became one of the most important methods of industrial organization.

Vertical integration resulted when a single organization or company controlled all stages in the conversion or transformation of raw materials to the final products and their dissemination. This included the sources of the raw material, such as an iron or copper mine or oil field, the places of production such as a factory or refinery, the means of distribution, and the marketing or selling of the final product. Andrew Carnegie's steel company is the best example of nineteenth-century vertical integration. The Bell Company in its early years, Alcoa, General Electric, the Ford Motor Company, Gustavus Swift (1830–1903) the Chicago meat packer, and James B. Duke (1856–1925) the North Carolina tobacco tycoon and founder of the American Tobacco Company in 1890, the oil companies, Exxon, British Petroleum (BP), and Royal Dutch Shell, also established vertically-integrated organizations.

Horizontal integration resulted when a single organization or company controlled one or possibly two of the stages in the manufacture of a product, such as the refineries in the petroleum industry which John D. Rockefeller controlled. Microsoft, media organizations such as the News Corporation (radio, television, newspaper, book publishing), and Time Warner (film, publishing, and television) are later examples of horizontal integration. AT&T represents a combined vertically and horizontally integrated company.

II. LEADING INDUSTRIAL ORGANIZERS

J. P. (Jay Pierpont) Morgan, Andrew Carnegie, and John D. Rockefeller were among the most prominent organizers. They committed themselves to the establishment of *order*, *efficiency*, and *stability* in the industrial world and claimed to believe in a free market, capitalistic economic system, while simultaneously trying to monopolize their industrial sector of the free market. Many of the industrial organizers had little interest in the political and social consequences of industrial development and were usually Social Darwinists.

J. P. Morgan (1837–1913), educated in Europe and at Harvard, was a financial investor, banker, and connoisseur of art in New York City. Morgan joined his father's firm in London in 1856 before returning to New York City in 1857 to work for George Peabody & Co. That same year he and Anthony J. Drexel (1826–93) formed Drexel, Morgan & Co. Known after 1895 as J. P. Morgan & Co., it became one of the world's most powerful banking houses.

Morgan was a business organizer and stabilizer who rose to prominence in 1873, the year in which a severe five-year panic began. His firm dominated government financing and managed a series of railroad and other corporate reorganizations, intending to restore international confidence in American securities. Morgan criticized competing railroad companies for using rebates and drawbacks, which were rate reduction techniques to attract customers and reduce competition. He worked to stabilize the railroad industry after the 1893 panic while simultaneously benefitting financially and concentrating power in his own hands.

Morgan believed in the gold standard and in 1895 established a syndicate that halted the gold drain from the reserves of the United States treasury. His intervention netted him a $6 million profit from the sale of government bonds. Morgan's formation of the overcapitalized but highly successful US Steel Corporation in Pittsburgh, in 1901 represented another bold undertaking. He also financed Edison, AT&T, and the General Electric Company. Morgan ruthlessly centralized control of industry and credit, helping stabilize corporate America but contributed nothing to social reform. A 1912 congressional investigation that claimed he represented a money oligarchy, controlling the nation's banks, corporations, railroads, insurance companies, and stock exchange, failed to tarnish his reputation.

Morgan gave his excellent personal library for use as a public reference library in New York City. A hypochondriac, owner of yachts and homes, he engaged in several affairs with fascinating women, including sharing a mistress with Edward VII (1841–1910) of England. The government established the Federal Reserve System in 1913 only months after his death in order to break up control of the eastern banking establishment that Morgan dominated.

Andrew Carnegie (1835–1919), self-proclaimed distributor of wealth for the improvement of mankind, emigrated with his family from Dunfermline, Scotland, to Allegheny, Pennsylvania, in 1848. He worked as a bobbin boy in a cotton factory earning $1.20 a week, then as a messenger and operator in a Pittsburgh telegraph office. In 1853 Thomas A. Scott (1823–81), later vice-president and president of the Pennsylvania Railroad Company, hired him as personal telegrapher and private secretary. Carnegie remained with Scott until 1865, during which time he introduced the use of Pullman sleeping cars, assisted with troop transportation during the Civil War, and organized the military telegraph department.

From 1865 to 1877 Carnegie succeeded in the expanding iron industry, in the oil business, and in selling railroad securities abroad before investing his fortune in the new American steel industry and establishing the Edgar Thomson Steel Works in 1875 in Braddock, Pennsylvania (Pittsburgh area). His strategy of "putting all his eggs in one basket and watching the basket" earned huge profits.

By 1889 American steel production surpassed British production and had become the world's leader.

Carnegie's organization ensured his success and included the engineer Captain Bill Jones (1839–89), Henry Clay Frick (1849–1919), and Charles M. Schwab (1862–1939). His company, until shortly before US Steel absorbed it in 1901, remained a limited partnership in which eight working associates, whose number increased to 40 by 1900, owned every share. Carnegie always held a majority 59 percent interest and distributed the remainder based on each partner's performance. He was a successful innovator, perfecting the use of *vertical integration*, and purchased the latest equipment even during times of depression and low costs. Carnegie's friends in the literary and political world included Herbert Spencer (1820–1903) the Social Darwinist and person to whom he said he owed the most, Theodore Roosevelt (1858–1919), Mark Twain, and the statesman and politician Elihu Root (1843–1937).

As the second richest American to date after Rockefeller, with a $298.3 billion fortune calculated in 2000s dollars, Carnegie in 1889 published in an article entitled "Wealth" in the *North American Review*, in which he described the responsibility of the wealthy to hold surplus wealth in trust for the public benefit. He believed the accumulation of great wealth required an exceptional person, obligated to employ his talents in distributing his fortune for the improvement of humanity. Carnegie's praise of the accumulators of great wealth as exceptional people represents the Social Darwinists affirmation of the nineteenth-century's *Gospel of Wealth* and persists in the twenty-first century.

By 1900 and now the richest man in the world, Carnegie prepared to apply his theory. He retired from the steel business, and in 1901 he sold Carnegie Steel to J. P. Morgan's US Steel Corporation for $482 million. US Steel, consisting of eight other companies, was the first billion dollar corporation with a value of $1.4 billion. The foundations, institutions, and agencies Carnegie established provided $350 million to support scientific research, erect public library buildings, advance teaching, promote international peace, and reward heroic acts and other worthy causes.

John D. Rockefeller (1839–1937) ranks as the United States' foremost industrialist and philanthropist. His father, who traded in lumber, salt, and other commodities, had farms in Richford, New York, and later in Moravia, New York. Rockefeller attended schools in New York and two years of high school in

Cleveland, Ohio, where the family moved in 1853. He worked for three-and-a-half years with a Cleveland company of commission merchants, leaving in 1859 to form a successful partnership selling grain, hay, meats, and other grocery products.

Rockefeller saw the commercial possibilities in Cleveland's recent railroad connection to the new oil regions in northwestern Pennsylvania, and in 1863 he and his partners built a refinery which in two years was Cleveland's largest. Rockefeller bought out his partners in 1865 and with his brother William (1841–1922), Samuel Andrews (an original partner), and another partner Henry M. Flagler (1830–1913), built a second refinery called the Standard Works.

Rockefeller's efficiency and unchecked tactics in the intensely competitive oil business soon produced an economically-effective operation. The company built its own barrels and warehouses, obtained fleets of freighters and tankers, and instituted an accurate cost accounting system. In 1870 the company became a joint-stock corporation, the Standard Oil Company of Ohio, capitalized at $1 million.

Rockefeller maintained that to eliminate destructive competition and achieve financial success required organization, and he began to acquire all of Cleveland's refineries and then refineries in other cities. By 1877 he controlled 90 percent of the country's refineries and had established a powerful *horizontally-integrated* organization. Rockefeller next acquired ownership or control of pipelines, purchased or leased oil-terminal facilities in New York City, and established an effective marketing system. Standard Oil now had a near monopoly in the American oil industry, and by the early 1880s it dominated the foreign oil market.

Rockefeller's monopolistic organization presented a problem. Ohio law clearly stipulated that any Ohio corporation, including Rockefeller's Standard joint-stock corporation with headquarters in Cleveland, had no legal right to own property or stock in states outside Ohio. Rockefeller's lawyer Samuel Dodd (1836–1907), having concluded a stockholders' trust agreement in 1879, proposed a solution in 1882. Standard established a board of trustees that took over all company stock and issued to the trustees trust certificates equivalent in value to the original stock certificates. The Standard Trust, which Rockefeller and his associates revealed in 1888 at a hearing before the New York State Senate in Albany, consisted of nine trustees (stockholders) and 41 companies in different states. This was the birth of the modern trust and the reason for the 1890

Sherman Antitrust Act directed at Standard Oil, Duke's American Tobacco Company, and other trusts.

Rockefeller claimed Standard never intended to obtain absolute control of refining, nevertheless, Standard's ruthless competitive practices offended and angered Americans who opposed monopoly. An Ohio court order in 1892 dissolved Standard of Ohio's trust agreement. Standard, however, controlled its management by interlocking directorates of its state companies and then by relocating in 1899 to Bayway, New Jersey, and establishing a holding company, Standard Oil of New Jersey.

Although the Sherman Antitrust Act did not consider a holding company an intrinsically illegal organization, Standard Oil was a monopoly that the Supreme Court broke up in 1911. At that time Standard, valued at $860 million, divested 33 of its affiliates and 57 percent of its worth. Standard Oil of New Jersey later became Exxon, several other companies of the original trust took on new names. Standard Oil of New York became Mobil, Standard Oil of Indiana became Amoco, and Standard Oil of California became Chevron. Only Standard Oil of Ohio, of all the original Standard Oil companies, retained the Standard Oil name.

After 1897 Rockefeller, like Carnegie, devoted himself to distributing his vast fortune and to establishing several philanthropic institutions. These included the Rockefeller Institute for Medical Research in New York City in 1901 (in 1965 Rockefeller University), the General Education Board of 1902, the Rockefeller Foundation in 1913 to promote the well-being of mankind throughout the world, and the Laura Spelman Rockefeller Memorial Foundation in 1918 and in 1929 part of the Rockefeller Foundation. Rockefeller's lifetime donations totaled $550 million. He remains the richest man in America with a fortune, estimated in 2000s dollars, of $318 billion. When Congress enacted an income tax in 1893, which the Supreme Court ruled illegal in 1896, Rockefeller already had a million dollar income. He earned $1,247,252.65 in 1894 compared to the $1,000 or less annual wage of most American workers. Rockefeller paid $14,961.39 in federal income tax at a rate of two percent after claiming deductions of $499,183.26. He filed and paid *under protest*.

Most of the industrial organizers such as Edison, Morgan, Carnegie, Rockefeller, Swift, and Duke were Social Darwinists and believed in laissez-faire economics. Working conditions in their companies were usually harsh. An unskilled worker in Carnegie's steel company, and in most other industries, earned 10¢ an hour and worked an 80-hour week, a skilled worker earned twice as much. Though none of them promoted social reform, most became philanthropists. Carnegie donated his $350 million to establish organizations such as the Carnegie Institute of Technology in Pittsburgh, Carnegie Corporation in New York City, and 1,679 Carnegie libraries in 1,412 American cities. Rockefeller's $550 million in philanthropy, in addition to establishing the Rockefeller Institute and Foundation, endowed the University of Chicago with $30 million in 1891. Duke provided the $60 million endowment for the establishment of Duke University in 1924, and Ford established the Ford Foundation in New York City in 1936. The benefactors controlled the membership and therefore the policies and objectives of their foundations.

III. ORIGINS OF INDUSTRIALISTS AND ENTREPRENEURS WHO DOMINATED AMERICAN INDUSTRY

Was it really rags to riches? No, many industrialists and entrepreneurs came from families that already had succeeded in business and were financially well off. Many had good educations. Carnegie and possibly Rockefeller represent the exceptions not the rule. Over 50 percent of these entrepreneurs, including J. P. Morgan, attended colleges, and 70 percent came from middle class families. About 48 percent of their fathers had business careers, and most of them shared the same white, Anglo-Saxon, Protestant heritage.

What drove them? The entrepreneurs wanted power, challenge, and adventure once they had acquired substantial wealth. Rockefeller Family Associates at one time in the 1960s–70s had a worth of $640 billion or about one-half the gross national product of the United States.

All of them had a passion for work and for details. Rockefeller's counting the number of drops of solder necessary to seal oil containers and the number of barrel bungs (stoppers) in the company's warehouses remains legendary. They all intensely disliked disorder and waste.

IV. REACTION TO THE GROWTH AND CONSOLIDATION OF INDUSTRY

Public reaction

Criticism of nineteenth-century American industrialism appeared in the writings of several well-known authors, such as Henry George, Edward Bellamy, and Henry Demarest Lloyd. They warned of government's and business's acceptance of laissez-faire economics and Social Darwinism's *survival of the fittest* which they said brought out the worst abuses of capitalism. Mark Twain feared that society in its adulation of every new technological innovation would fail to see all the consequences of its own creations.

Henry George (1839–97), a San Francisco newspaperman who later moved to New York City and ran for mayor, criticized the increasing industrial society in his 1890 economic treatise *Progress and Poverty*. He raised the question that if the nineteenth-century industrialism represented progress, then why did so much poverty exist in the industrial cities? Herbert Spencer, the English Social Darwinist and champion of individualism and laissez-faire, was a favorite target. George said Spencer was so blind to the realities of life that he would call a race between two swimmers, one loaded with cork and the other with lead, a fair race.

George also criticized the English political economist and minister Thomas Malthus (1766–1834) who argued in his *Essay on the Principle of Population* (1798) that population always increased faster (geometrically) than the food supply (arithmetically) and that society should therefore let natural disasters, war, and famine take their toll. The Social Darwinists and laissez-faire economists were completely wrong according to George. George's book sold very well, at least 700,000 copies, considering that he wrote a serious economic treatise and that it underwent numerous foreign language translations. George died, apparently of a stroke, during his New York City mayoral campaign of 1897.

Edward Bellamy (1850–98), son of a Baptist minister and a socialist utopian author in Springfield, Massachusetts, published his utopian novel *Looking Backward 2000–1887*, in 1888. In this well-known criticism of the abuses of nineteenth-century, capitalistic United States, the hero, Julian West, falls asleep in 1887 and awakens in 2000. During the interval, cooperation has replaced competition and selflessness replaced selfishness. This was Bellamy's vision for a future United States. Bellamy's popular book sold about 200,000 copies by 1890.

Henry Demarest Lloyd (1847–1903), a Dutch Reformed minister's son, journalist lawyer, and champion of labor in Chicago, published *Wealth Against Commonwealth* in 1894. Lloyd asked whether society should strive for the wealth of a few or the wealth of society, meaning the common wealth. He directed his attack against monopolies, particularly Rockefeller and Standard Oil. Lloyd was one of the first muckrakers, a description applied to nineteenth-century moralistic, investigative reporters. He joined the Socialist Party in 1903.

Mark Twain (1835–1910), author and humorist and one of the country's best known writers who later resided in Hartford, Connecticut, published his technological critique *A Connecticut Yankee in King Arthur's Court* in 1889. In this novel Hank Morgan, a nineteenth-century Yankee, arrives in medieval England, bringing with him democracy and all the latest technological advances including weapons, such as machine guns and bombs. A war breaks out, and the technologically advanced Yankees defeat the Knights. Twain, at this point in the story, seemed to praise the new technology. After a peaceful interval another war breaks out, and the Yankees, now armed with even more advanced weapons such as Gatling guns and electrified fences, defeat the Knights a second time.

Once again Twain seemed to praise the advance of technology. But the Yankees in their defense against the Knights had fortified themselves in a bunker, and the bodies of the 25,000 Knights they killed have trapped them in their bunker. Unable to escape, the germs released by the decaying bodies have doomed the entrapped Yankees, and soon all have died from disease. Technology has turned on its users; the victors have become the victim. Twain's message then, and which remains meaningful, is that if society's technology makers fail to control their technological creations, technology will destroy its makers.

Industrial leaders, such as the Social Darwinists Carnegie and Rockefeller, preached the *Gospel of Wealth.* They argued that concentration of wealth was necessary for the good of society and that the holders of wealth were special individuals who had the skills and knowledge to deal with society's ills. Only they knew best how to use their wealth to produce the most beneficial social results.

Political reaction from state and federal governments: regulation of railroads and corporations

Regulation of railroads

Oliver Kelley (1826–1913), a Minnesota farmer, founded the Grange movement in Washington, DC, in 1867. The Granges initially were social-cultural societies of farmers and rural inhabitants but became political activists and turned to the courts to help end the abuses the railroads inflicted on farmers. The first of two important Supreme Court cases in which the Grange challenged existing railroad legislation was *Munn vs. Illinois* in 1877. It dealt with the exorbitant rates the railroads charged farmers for grain elevator storage and resulted in a victory for the Grange and the states over the railroads.

In the second Supreme Court case, *Wabash vs. Illinois* in 1886, the railroads were appealing a state ruling that forbade railroads from charging more for a short haul than for a long haul when both hauls involved interstate travel. The court ruled that the state of Illinois could not regulate rates on interstate travel between Illinois and any other state, such as Wisconsin, Iowa, Indiana, Missouri, or Kentucky, and was a defeat for the Grange and the states. Because the case involved regulation of interstate travel, an individual state could not resolve the problem and clearly demonstrated that the country needed a federal regulation to deal effectively with interstate travel. The 1824 *Ogden vs. Gibbons* Supreme Court case revealed the same lack of federal legislation.

Consequently Congress passed the Interstate Commerce Act in 1887 to solve the railroad problems. The act prohibited discriminatory rates, railroad pooling, rebates, and drawbacks. Nor could railroads charge more for a short haul than for a long haul over the same line involving interstate travel. Republican Senators William Windom (1827–91) of Minnesota in 1874 and especially Senator Shelby Cullom (1829–1914) of Illinois in 1885 were instrumental in promoting regulations that became the 1887 Interstate Commerce Act.

Because no branch of government enforced the 1887 Interstate Commerce Act during nearly 20 years of laissez-faire economic philosophy, Congress in 1906 during the Progressive period and Theodore Roosevelt's presidency passed the Hepburn Act to strengthen the weak 1887 act. The Interstate Commerce Commission (ICC) now set railroad rates rather than merely approving them as it had done under the Interstate Commerce Act.

Regulation of corporations

The Sherman Antitrust Act of 1890, named after Republican Senator John Sherman (1823–1900) of Ohio, was a response to the abuses of big industry. According to its provisions, Congress declared illegal all trusts, combinations, cartels, and conspiracies in restraint of trade, but not holding companies. Like the Interstate Commerce Act, the government did not vigorously enforce the Sherman Antitrust Act until 1901 under presidents Theodore Roosevelt and William Howard Taft (1857–1903). Consequently the courts heard only 14 antitrust cases from 1890 to 1901.

Congress passed the Clayton Antitrust Act in 1914 to strengthen the 1890 antitrust act. The 1914 act forbade the courts from treating labor as a commodity, as it had done in a previous ruling that led to the six-month imprisonment in Woodstock, Illinois, of American Railway Union leader Eugene Debs (1855–1926) for restraining trade during the 1894 Pullman Railway strike in Chicago.

Alternatively, the Supreme Court in an eight to one decision in the 1895 case *United States vs. E. C. Knight Company* found the American Sugar Refining Company, of which E. C. Knight Company gained control in 1892 and thereby controlled 98 percent of the sugar industry, not in restraint of trade. The Supreme Court declared that manufacturing (refining) was a local decision not subject to congressional regulation of interstate commerce.

Other reactions to industrial growth occurring at this time included the rise of labor unions, immigration, rise of cities, and the emergence of reformers critical of the abuses of unrestrained industrialists.

CHAPTER 4

ENVIRONMENTAL POLLUTION: THE OTHER SIDE OF URBANIZATION, MODERNIZATION, AND INDUSTRIAL DEVELOPMENT

I. INTRODUCTION

Where there is water, there is life as we know it. Society accepts this statement as a self-evident truth or a law of nature, and it may be the closest society has come to proclaiming an absolute truth. Similarly, the statement, where there is air, there is life as we know it, has become another truth that few would challenge. The historical record reveals that the earliest societies recognized the importance of clean water and clean air to their well-being. Yet the same historical record shows their disregard and lack of concern for water and air quality. The historical record clearly is mixed on whether to pollute, whereas it shows quite clearly that from earliest times society always seemed more concerned about water pollution than air pollution.

Land pollution received the least concern and appeared more an aesthetic issue than a health issue. Destruction of forests by fire, depletion of soil, making it unfit for agriculture, were more serious concerns rather than pollution of land by industrial waste deposits until the twentieth century and the production of non-degradable plastics. Society generally believed the discarded products, garbage, animal remains, and other waste eventually would decay or disintegrate.

Even with regard to water pollution the primary concern was the quantity of water available rather than the quality of the available water. Consequently societies, such as the Romans, who from 312 BC to 455 AD, constructed extensive aqueduct systems that transported throughout their empire an estimated

130 million gallons of water daily, achieved little understanding of water quality and left a sparse record.

The earliest records, nevertheless, indicate that the aesthetic qualities of water, its appearance (cloudiness or turbidity), taste, and smell, were the problems the first water treatment processes tried to solve. Thousands of years elapsed before society recognized that the senses alone were an insufficient judge of water quality. Though neither water nor air pollution received serious attention until the later nineteenth and early twentieth centuries, society has continued to see water pollution as more threatening and has improved the quality and quantity of its water resources. Holding back progress in water and especially in air pollution regulations, in addition to the discredited economic argument about high costs, was an old but still widely held erroneous belief in the ability of water and air to dilute indefinitely. Pass the air pollutants into the voluminous atmosphere and water pollutants into rivers, lakes, or oceans and the air and water will carry away the pollutants. Somehow, perhaps miraculously, dilution has enabled water and air to diminish or eliminate the harmful effects of every impurity discharged within them.

The indisputable scientific link between water pollution and disease in humans in the second half of the nineteenth century challenged society to improve its water quality. The microscope beneath the eyes of Filippo Pacini, Carl Eberth, Robert Koch, and others provided the evidence. But no unchallenged scientific evidence linked air pollution and disease, despite the obvious effects of pollution on the aesthetic quality of the atmosphere, the environment, and the health of inhabitants breathing it. Scientists had established indisputable quantitative chemical tests for water pollutants by the mid-twentieth century. Unfortunately, many of the chemical tests providing quantitative evidence for air pollution that the polluters and their defenders demanded (and which electronic, gravimetric, and electrochemical instrumentation; colorimeters, spectroscopes, and gas chromatographs provided), had not advanced sufficiently until the mid-twentieth century. Spectroscopy for metals began in the 1940s–50s, chromatography for purgeable organics only in 1953.

In the United States environmental pollution emerged as a serious but too often neglected problem in the nineteenth century. Water pollution received much more attention and much more public spending on its prevention than air pollution. The reason is clear. No one would drink or bathe in polluted water, especially when the pollutants were clearly visible. Clean water was absolutely essential for life. American society expected its scientists and engineers to provide clean water sources. Pollutants in the air, except when highly concentrated and visible because of their color or odor, seemed less likely to have serious health consequences, and American society generally ignored air pollutants. Most individuals believed that the atmosphere would dilute the pollutants sufficiently and render them harmless. Smokestacks would pass the pollutants high into the atmosphere and thereby do little damage.

Water purification in the United States, and in many other countries in Europe, Asia, and elsewhere, went through three stages. The first stage required the construction of water distribution and sewer systems. Stage two introduced slow sand and rapid sand filtration systems and the addition of coagulants. The third stage saw the addition of chlorine gas or hypochlorite compounds to kill deadly bacteria and disinfect water. Its technological development, like most technological innovations, required the expertise of scientists and engineers to provide society with clean water.

The technological development of water purification therefore provides a good example of the science-engineering-society interplay. Science discovered the causes of the pollution; engineering developed or produced the instruments, implements, or tools that reduced or eliminated the pollution; and society benefitted from the science-engineering breakthroughs that resulted in a healthier and safer environment.

Air and land pollution experienced no similar nineteenth-century technological advancements.

II. WATER PURIFICATION AND THE PROBLEM OF WASTE DISPOSAL IN THE UNITED STATES: THE THREE STAGES OF DEVELOPMENT

Background

For centuries physicians, scientists, engineers, and other concerned citizenry recognized the obvious connection between clean drinking water and good health. They boiled and filtered (strained) their drinking water to improve its appearance and taste and would not drink, or bathe, in dirty or visibly polluted water. The earliest known records on treating water to improve the aesthetic qualities of taste and smell date from 2000–1500 BC. Sanskrit writings record the use of filtration through sand, gravel, or charcoal, boiling, and aeration (exposure to sunlight) as recommended methods. The Egyptians recognized a link between the cloudiness of water and an unpleasant taste and appearance, and as early as 1500 BC they added alum, a known mordant or dye-fixer, to sink and settle the suspended particles. The ruins of Cretan and Assyrian cities dating from 1500–1400 BC indicate the existence of sanitary sewers. Hippocrates (460–377 BC) in Cos recommended boiling and straining rainwater to prevent it from having a bad smell and causing hoarseness. Aristotle (384-322 BC) in Athens cautioned Alexander the Great (356-323 BC) not to let his men drink from stagnant pools but to carry boiled water with them. Centuries later, in the 1200s, in Italy's medieval Ferrara, the city passed a law that limited the number of butcher shops in an area in order to control the waste (blood and entrails) butchers discharged into the Po River and on the land.

During the 1700s filtration became an effective technique for removing particles from water. The Italian anatomist Luc Antonio Porzio (1639–1723) in Padua in 1685 published the first known illustration of a filtration system that included sedimentation, straining, and sand filtration. It had three filtration units, each having two compartments, one with a downward flow filter and one with an upward flow filter. Porzio designed his system to provide soldiers with safe drinking water as a result of the sanitation problems he observed during the Austrian-Turkish war of 1685. In 1746 the Parisian scientist Joseph Amy (1697–1760) received a patent for a sponge, charcoal, and wood filtration system that he sold to homes, and in 1791 the English architect James Peacock (1738–1814) in London, obtained the first British patent on a filtration system. His three-tank system filtered water by passing it through graded layers of sand and gravel.

Up to this time, no one connected unpleasant smelling and tasting water sources and the often-fatal intestinal diseases later identified as cholera, typhoid fever, and dysentery. Variations of the miasma, or bad air, theory of disease, particles emanating from rotting organic matter causing the bad air, and even divine intervention received greater acceptability until the mid-1800s investigations of John Snow (1813–58). Snow, a vegetarian, teetotaling-practicing physician in London, survived three cholera epidemics during which time London's population grew from 1.95 million in 1841 to 3.2 million in 1860.

Snow investigated the cause of a cholera epidemic that struck England, Europe, and North America in 1831 for the first time. In 1849, after the 1848–49 cholera epidemic killed 14,600 Londoners, Snow in a short pamphlet questioned the prevailing miasmic theories and proposed instead a connection between cholera, a disease that killed with terrifying quickness in as little as eight hours, and a contaminated water supply. In his definitive study carried out during London's 1854 cholera epidemic in which another 10,675 people died, Snow used microbiological and statistical evidence to prove that drinking water pumped from the Thames River contaminated with animal and human feces had killed 600 people in only 2–3 days. Snow traced the source to polluted drinking water obtained from a pump on the corner of Broad Street and Cambridge Street in London's Soho district. The well lay only three feet from an old cesspit originally located under a nearby house.

Few people, including the London Board of Health, at first believed Snow. The Reverend Harry Whitehead (1825–96), vicar of St. Luke's Anglican Church, claimed divine intervention had caused the cholera outbreak. The medical establishment generally accepted the miasma theory that poisonous vapors released from decaying plant and animal matter into the air caused the disease. Ironically, in 1854 the Italian physician Filippo Pacini (1812–83) at the University of Florence, a believer in the germ theory of disease, discovered the short comma-shaped cholera bacterium (*Vibrio cholerae*) during the 1854–55 outbreak in Florence. He performed autopsies on cholera victims and with his microscope observed the cholera bacterium in their intestinal mucosa (mucous membrane). The medical community largely ignored his work and attributed the disease to miasma. Pacini, who died poor, received belated recognition for his discovery, culminating in 1965 when the medical profession's International Committee on Nomenclature named the cholera bacterium *Vibrio cholerae Pacini 1854* in his honor.

Fortunately, advances in bacteriology enlightened medical opinion in the next decade. First, Louis Pasteur (1822–95) in Paris beginning in 1861 introduced his germ theory of disease. Then, in 1880 the German physician Carl Joseph Eberth (1835–1921) at Zurich discovered the rod-shaped bacterium responsible for typhoid fever (*Salmonella typhi bacillus*) in the spleen and mesenteric lymph nodes of victims. In 1883 another German physician Robert Koch (1843–1910) at the Health Office in Berlin rediscovered the bacterium responsible for cholera in the intestinal mucosa of cholera victims, thereby silencing Pacini's critics and skeptics. Koch's discovery was the latest of his significant medical breakthroughs.

Already in 1881 in a 48-page paper Koch demonstrated the deadly effect of a calcium hypochlorite solution on bacterial cultures. In 1882 he isolated the rod-shaped bacterium that caused tuberculosis (*Tubercle bacillus*), likely his greatest discovery, and in 1884 his co-worker, the German bacteriologist George Gaffky (1850–1918), first isolated the rod-shaped typhoid fever bacterium from a victim's spleen, confirming Eberth's discovery.

Dysentery, the third member of the unholy trinity of waterborne diseases, is an infection of the large intestine. The cause is either a protozoon *Entamoeba histolytica* (amoebic dysentery) or bacteria of the genus *Shigella* (bacillary dysentery). Friedrich K. Lösch (1840–75) at St. Petersburg in 1875 clearly recognized that an intestinal protozoon (parasite) caused amoebic dysentery. Confirmation came in 1903 in the tropical disease studies of the German zoologist Fritz Schaudinn (1871–1906) at Berlin. Schaudinn, in tests on himself, demonstrated that *Entamoeba histolytica* was the pathogen of amoebic dysentery and not the harmless *Entamoeba coli*. He named both protozoa.

Discovery of the pathogen responsible for bacillary dysentery, or bloody diarrhea, occurred in 1897. The Japanese bacteriologist Kiyoshi Shiga (1871–1957), then at the Institute for Infectious Diseases in Tokyo, isolated the rod-shaped dysentery bacillus from tissue in the large intestine. Shiga named the bacterium *Bacillus dysenterie* (in 1930 renamed *Shigella dysenteriae*). Confirmation of Shiga's discovery came three years later when Simon Flexner (1863–1946), then at Pennsylvania, isolated the dysentery bacillus.

THE THREE DEADLY WATERBORNE DISEASES			
Disease	*Discoverer*	*Year*	*Effect on Humans*
Cholera	John Snow in London established the link between contaminated water and cholera	1854	Abdominal pain and watery diarrhea, rapid dehyzdration and shock
	Filippo Pacini in Florence discovered the cholera bacterium (*Vibrio cholera*)	1854–55	
	Robert Koch in Berlin isolated the cholera bacterium	1883	
Typhoid Fever	Carl Joseph Eberth in Zurich discovered the bacterium *Salmonella typhi bacillus*	1880	Fever, rash, spleen and lymph node enlargement, gastrointestinal bleeding and ulceration
	George Gaffky in Berlin isolated the typhoid bacterium	1884	
Dysentery (amoebic)	Friedrich Lösch in St. Petersburg recognized that the parasite/protozoon *Entomoeba histolytica* caused dysentery	1875	Diarrhea and constipation accompanied by mucus and sometimes blood
	Fritz Schaudin at Berlin confirmed Lösch's work	1903	
Dysentery (bacillus)	Kyoshi Shiga in Tokyo isolated the bacterium *Bacillus dysenterie* (*Shigella dysenteriae*)	1897	Abdominal pain, diarrhea, and fever associated with poor hygiene
	Simon Flexner at Pennsylvania isolated the dysentery bacillus	1900	

As a result of these advances in bacteriology and the increasing acceptance that water polluted with human and animal feces, animal parts, sewage, and other contaminants caused cholera, typhoid fever, and dysentery, cities in Europe and the United States in the nineteenth century took three important steps to purify their water supply systems.

First came the construction of water distribution systems (waterworks) consisting of pumping stations and underground pipes to bring water from nearby rivers, lakes, or bays into the cities and underground sewer systems to carry the raw sewage and other waste matter from the cities into the rivers, lakes, or bays far from the pumping stations. Air and sunlight (aeration) decomposed some of the undissolved solids in the polluted water. Sand filtration and then coagulants with sand filtration units that removed undissolved solids and considerable bacteria, viruses, and protozoa (pathogens) followed. Finally in the first decades of the twentieth century, city after city began adding chlorine to its drinking water sources killing nearly all the pathogens and resulting in the elimination of waterborne diseases. Taking into account that a human on average drops 3.59 ounces of solid stool a day (possibly 91 pounds a year), that a horse, of which about 3 million lived in American cities, can deposit as much as 20 pounds of manure per day and that some of the waste found its way into city water supplies, the public expected water purification.

Water distribution and sewer systems

The first advance in the water purification process was the construction of water distribution and sewer systems. Early city water distribution and sewer systems included those in Cleveland, Ohio, and Chicago, Illinois. The growing use of significantly improved flush toilets (water closet) in the larger cities required the construction of sewer systems that varied in size and efficiency. Brooklyn, New York; Boston and Lynn, Massachusetts; and Memphis, Tennessee, constructed some of the earliest sewer systems. The Cleveland and Chicago systems are of particular significance because of their size, innovation, and cost.

Cleveland, a city founded in a swampy malaria-plagued area on the shores of Lake Erie and the Cuyahoga River, had a population of 57 in 1810 and backyard water wells. A resident, Benhu Johnson, provided the city's first commercial water supply, delivering two barrels holding 50 gallons of Lake Erie water for 25 cents. As Cleveland's population grew, largely because of the Ohio-Erie Canal's construction, reaching 17,000 by 1840, it required more water, and in 1833 Philo

Scovill (1791–1875), a hotel owner, real estate businessman, and his associates organized the Cleveland Water Company to construct a water works to serve the entire community.

Scovill lacked the necessary resources to carry out the project. The city council, however, recognized the importance of an adequate clean water supply, and in 1840 it spent $35 to dig a public well in Public Square, followed by a network of wells and cisterns in the next decade. This proved inadequate, and when no entrepreneurs were willing to finance the cost of an adequate centralized water distribution system the city council in 1853, acting on its own proposal, *Report on the Subject of Water Works* (24 pp., 28 February 1853), authorized an expenditure of $380,766.65 to construct a water works.

The report's author, Theodore R. Scowden (1815–81), a Cincinnati hydraulic engineer who became Cleveland's chief engineer, designed and constructed a water works that had an aqueduct, or delivery pipe, 40 inches in diameter and 300 feet long. It brought water from Lake Erie to a pump house located as far from the Cuyahoga River as required to avoid city drainage and the river's discharges. The system's single Cornish steam engine (and an identical back up engine) could supply 3 million gallons of water per day (30 gallons per 100,000 inhabitants) to the 11 miles of wrought iron distribution pipes leading from the reservoir, although in 1856 it actually delivered about 38,000 gallons daily. Water delivery increased to 348,664 gallons per day in 1857 and to 5.5 million gallons per day in 1900.

Cleveland constructed its first sewer in 1856 and in 1858 began constructing a rudimentary sewer system that consisted of open drains to convey wastewater downhill toward the Cuyahoga River and Lake Erie. The resulting increase in pollution of Lake Erie's shoreline led to the construction in 1874 of a new water-intake crib and tunnel 6,600 feet offshore. The most expensive components of the water works (water distribution and sewer systems) were the 144-horsepower Cornish engines and boilers, each costing $32,500.00; the reservoir, $29,325.75; and the real estate on which to construct the engine house and reservoir tower, $30,000.00.

The operating cost per day was $18.89 and compared very well with steam engines operating at the Pittsburgh Water Works, which if used in Cleveland would have cost $33.24 per day. Annual operating expenses for the 3 million gallon per day system were $6,896.00. They included the coal consumed at the rate of 3 pounds per hour for each horsepower used in pumping water for 16 hours per day, 365 days per year (or 32,060 bushels at 10 cents a bushel) for a total of $3,206.00; the crew (chief engineer $2.50 per day, assistant $1.50 per day, two firemen each at $1.00 per day, laborer at $1.00 per day) and repairs and oils, $1,500.00.

The final cost of the 11-mile water distribution system when it began operating on 25 September 1856 was $527,000, the increase resulting from revisions to the original plan, particularly engineering difficulties, such as the unanticipated presence of quicksand. To pay for the water system the city in 1859 charged a one-time fee of $3.00 for a plumber to connect customers to the system and varying annual rates payable in two installments for homeowners, churches, business, and other consumers.

In a house (not exceeding five rooms) the annual rate was $5.00 plus 50 cents for each additional room, $2.00 for each bath tub, and from $2.00 to $5.00 for a water closet. A hotel and boarding house paid $1.00 per room, saloon and eating houses $5.00 to $25.00, a church $5.00 to $10.00, and a printing office $5.00 to $10.00. Cleveland's population more than doubled from 17,034 to 43,417 in the decade 1850–60, resulting in large increases in water consumption.

Chicago, situated on Lake Michigan and the Chicago River which divides the city into three parts, received a charter in August 1833. Its population of 350 residents grew rapidly to 2,000 residents the next year and to 4,500 in 1840. Its growth, however, led to the river's rapid pollution and to the city trustees' construction of a public well from which residents carried lake water in buckets to their homes and peddlers transported water in mule-drawn carts, selling it door-to-door for 10 cents a barrel.

With increasing growth and the need for better water distribution, the privately-owned Chicago Hydraulic Company, to which the Illinois legislature granted a 70-year charter, in 1842 constructed a $24,000 distribution system with a pumping station and several thousand feet of wooden water pipes beneath the streets. The system's delivery or intake pipe extended 150 feet into Lake Michigan, the city's abundant water source. A 25-horsepower steam-driven pump located in a station on Michigan Avenue pumped water to an elevated wooden standpipe tank from which it flowed by gravity through the wooden water pipes. The system's main problems included fish-clogged water intakes, turbid water after storms, and ice in winter.

The city of Chicago purchased the Chicago Hydraulic Company in 1852, after the State of Illinois in 1851 authorized the city to establish its own water works. By 1861 and the beginning of the Civil War, Chicago's water system had a 600-foot wooden intake pipe extending from Lake Michigan to the pumping station, an elevated standpipe tank (a vertical pipe to provide uniform pressure), a 95-mile distribution system of cast iron pipes, and three elevated wrought iron reservoirs each of 500,000 gallons capacity. Average daily pumpage was 4.8 million gallons of water for Chicago's 120,000 residents.

Chicago faced a major challenge at this time. Its sewer system through which flowed raw sewage and wastes from slaughter houses, distilleries, and other industries located on the banks of the Chicago River, had turned the river, which was the city's main highway for moving goods and services, into a cesspool. When it rained, sewage and waste reached Lake Michigan, contaminating water along the shoreline and the water intake cribs (collection points) located offshore. Because of the polluted river water flowing into the lake, Chicago's residents constantly suffered from outbreaks of cholera, typhoid fever, and dysentery. An 1849 cholera epidemic killed 678 (2.9 percent) of Chicago's 23,380 residents, the 1854 epidemic killed 3,600 (5.5 percent) of its 65,872 residents. Annual deaths from cholera averaged 65 per 100,000 population from 1860–1900, typhoid deaths reached 568 in 1881 and a high of 1,997 (174 per 100,000 population) in 1891. These numbers forced Chicago to recognize that it had a serious clean water supply problem and by 1900 led Chicago to complete three significant engineering projects that achieved international recognition.

In 1867 Chicago constructed an under-the-lake tunnel to deliver water from a wooden intake crib located two miles from the Lake Michigan shoreline to the Chicago Avenue Pumping Station. Ellis S. Chesbrough (1813–86), an engineer and since 1861 head of the Bureau of Public Works, directed the tunnel's construction. Dug through clay 60 feet under the lake's level and lined with two layers of brick, the five-foot diameter tunnel and the pumping station represented the beginning of modern water works with their intake cribs, tunnels, pumping stations, and filtration and purification plants. In 1869 the city completed construction of the still-existing Chicago Avenue Pumping Station and the Chicago Water Tower, the only building to survive the Great Chicago Fire of 8–9 October 1871 undamaged. (The Chicago fire swept through 3.5 square miles killing 250 and destroying 17,450 buildings in 27 hours.) Two years later, in 1871, the city explored the bold idea of reversing the flow of the Chicago River from eastward to westward to carry wastes away from Lake Michigan and down the Mississippi River.

Rudolph Hering (1847–1923) in Philadelphia, who became the chief engineer of Chicago's Drainage and Water Supply Commission in 1886 at a salary not to exceed $10,000, pointed out that a low point in the regional continental divide, an 8-foot high summit or ridge located 12 miles west of the lake shore, separated the Great Lakes drainage system from the Mississippi River drainage system. The plan that emerged in 1887 required engineers to cut through the ridge with a canal extending from the southern tip of the Chicago River and carrying wastes westward away from Lake Michigan and down to the Mississippi River through the Chicago, Des Plaines, and Illinois rivers.

The state legislature in 1889 passed a law that created the Metropolitan Sanitary District of Greater Chicago to execute the plan which it completed in 1900. The 28-mile lock-style canal, 24 feet deep and 160 feet wide, called the Sanitary and Ship Canal, or Main Channel, was the first of three canals Chicago constructed to divert wastes westward away from Lake Michigan and into drainage systems flowing to the west. The second, the North Shore channel, was an 8-mile long canal completed in 1910, and the third, called Calumet-Sag, was a 16-mile canal completed in 1922. The reversal of the Chicago River was the largest municipal earth project ever done at that time. The construction of the Sanitary and Ship Canal cost $33.5 million, and three canals cost $70 million. The entire system consists of 71 miles of canals, channels, and rivers.

The sewer systems constructed in other American cities varied in size and efficiency. Brooklyn's system, constructed in 1857–59 and where James Kirkwood (1807–77) and Julius Adams (b. 1812) were the city's only sanitary engineers, drained rainfall that fell at the rate of one inch per hour. Boston had 100 miles of city sewers in 1860, increasing to 226 miles in 1885. Lynn constructed a sewer system beginning in 1866 that in 1880–90 expanded to include 140 miles of pipelines. Memphis, following an 1878 yellow fever epidemic, in 1880 spent $137,000 on a dual twenty-mile system for the separate removal of sewage and storm water.

The Newport, Rhode Island sanitary engineer and former Union army major George Waring Jr. (1839–98) designed the Memphis system. It did not function effectively, clogging frequently because of the small diameter size pipe that connected houses to the sewer line. Waring insisted on 4-inch diameter pipe instead of the recommended 6-inch or larger diameter size. Waring accepted the miasma theory of disease. His death years later resulted from yellow fever, a mosquito-transmitted disease he contracted while investigating a yellow fever epidemic in Havana, Cuba.

The pipes used to construct Cleveland's, Chicago's, and other urban water distribution and sewer systems in cities such as Brooklyn, Boston, Lynn, and Memphis were usually cast or wrought iron (sometimes wood, clay, or brick), often 18–20 feet long. Diameters, depending on use, ranged from 4–6 inches, 12–16 inches, to 30–50 inches and cost about $47 per ton. Steam-powered pumps that moved water through the systems delivered anywhere from 270 to 1,150 horsepower. A 1,150 horsepower steam engine stood 60 feet high and weighed 850 tons.

The rapid industrialization and urban population growth in the decades after the Civil War accelerated the construction of water distribution systems and use of piped water in the home. By that time the 16 largest cities had constructed waterworks. By the 1890s–early 1900s the addition of home bathrooms with sinks and bathtubs, and especially the significantly improved flush toilets increased the need for sewer systems to deal with the problems of waste disposal and the water pollution it caused.

The earliest flush toilet in Europe dates from 1596 when Sir John Harrington (1561–1612), a London writer and Queen Elizabeth I's godson, built toilets for himself and for the queen. His toilet, named *Ajax*, consisted of a seat, a bowl, and a cistern (water tank) located behind the bowl and seat for removing the waste with running water.

Almost 200 years passed before Alexander Cummings (1733–1814), an Edinburgh-born London watchmaker, patented an improved flush toilet in 1775. Cummings' toilet contained a sliding valve between the toilet bowl and the trap (a u-shaped joint) and pipes that carried waste to the sewer. Water contained in the low part of the trap prevented bad odors from passing back through the toilet and into the surroundings. In later toilet designs the bad odors escaped through a pipe in the roof. Cummings' toilet cost 6 shillings 8 pence ($1.25 or £ 25.36 = $38.55 in 2002) to construct.

From 1861 Thomas Crapper (1836–1910), a London plumber, improved the flush toilet. He perfected the existing toilets' pull chain and cistern, which functioned to discharge sufficient water at high speed to cleanse the bowl. Contrary to popular legend Crapper, who had nine toilet patents, did not invent the flush toilet. Until 1885, when Crapper's partner, the London pottery maker Thomas Twyford (1849–1921), built the first one-piece, all-porcelain, trapless toilet, toilets were of wood and metal.

In the United States several inventors contributed to the perfection of the flush toilet. They included James T. Henry and William Campbell in Philadelphia who, in 1857, received the first American patent for a plunger type flush toilet and Waring in Newport. In 1882, Waring transformed the jet siphon flush toilet that William Smith in San Francisco patented in 1877 into a full-size porcelain toilet. Robert Frame and Charles Neff in Newport and Fred Adee in New York City made additional improvements. (Henry and Campbell, US Patent 18,972 of 29 December 1857; Smith, US Patent 190,919 of 15 May 1877; Waring, US Patent 266,404 of 24 October 1882; Frame and Neff, US Patent 425,416 of 15 April 1890; Adee, US Patent 436,451 of 16 September 1890).

The Dececo Company in Boston marketed the improved toilet in 1890 as the Dececo siphon water closet. Its features, such as its large area of standing water and self-siphoning action, enabled Dececo to sell a significant number of toilets. In Washington, DC, most of the recently-constructed first-class hotels, and office buildings, including the Capitol Building, had Dececos. Because of these improvements, the number of toilets in American cities increased markedly. Boston in 1860 had 6,500 flush toilets, 100,000 in 1885. In 1880, 25 percent of households in Chicago, population 503,185, had toilets. The price of a toilet varied widely. A three-piece bathroom set consisting of a tub, lavatory (sink), and toilet that the J. L. Mott Iron Works in New York City listed in its 1898 *Trade Catalogue* cost anywhere from $200 to $1,200. Standard Sanitary Manufacturing Company in Pittsburgh gave a price of $69.75 for a bath, lavatory, and toilet. The Fred Adee Company in New York City sold its siphon toilet for $41.00 in 1898.

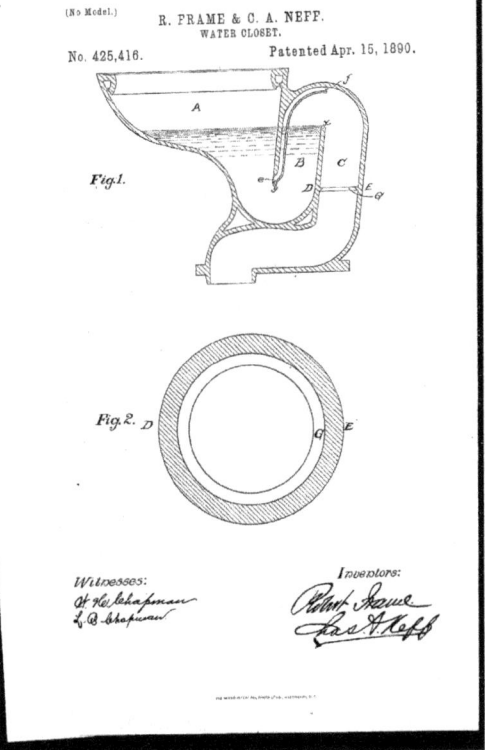

Figure 29 Siphon Water Closet

Toilet paper appeared at this time. Joseph Gayetty (b. 1827) in New York City introduced Gayetty's medicated, or therapeutic, paper in 1857. Each sheet had his name watermarked (imprinted) on it and contained aloe, a healing plant extract. A package of 1,000 sheets sold for $1.00. Gayetty's toilet paper replaced newspapers, catalogue pages, and leaves. Twenty years later, in 1880, the British Perforated Paper Company patented and sold toilet paper in boxes containing small pre-cut squares. The brothers Thomas, Edward (1846–1931), and Clarence Scott, who in 1867 established the Scott Paper Company in Philadelphia, introduced toilet paper on a roll in 1890. By that time Scott had become the country's leading toilet paper producer.

Slow and rapid sand filtration systems

The second advance in water purification was the introduction of filtration systems in which water passed either slowly (slow filtration) or rapidly (rapid or mechanical filtration) through or over fine gravel or sand. Slow sand filtration, which in its simplest application dates from 2000–1500 BC and in improved design from the early nineteenth-century in Europe, removed cloudiness, undissolved particles, and deadly pathogens from water. By 1900 scientists in Europe and the United States had proved its effectiveness in reducing outbreaks of cholera and typhoid fever in cities that filtered their water compared to those cities that failed to filter their water.

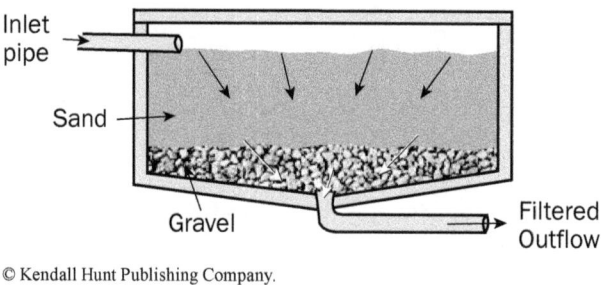

© Kendall Hunt Publishing Company.

Figure 30   Sand Filtration

In slow sand filtration, water passed down through a layer 1.0–1.5 meters or more of washed, uniform, fine sand, having small pore spaces (0.15-.035 millimeters) between the sand grains, at the rate of 100–200 liters per hour per square meter (3 to 6 gallons per hour per square foot) of sand area. It effectively treated low-turbidity small water sources such as well or spring water by removing clay and pathogens (polio virus, cholera, typhoid fever, and dysentery bacteria). The top one to three millimeters of sand, which after one or two days formed a sticky mat (schmutzdecke), contained biologically active microorganisms, such as predatory bacteria of the genus *pseudomonas* and trichoderma that came from the water. The microorganisms broke down organic matter and eliminated 91–99 percent of the pathogens. The sand filter bed eliminated any suspended inorganic or organic matter by straining. Plant workers regularly removed the top layer of sand to clean the filtration system, and after a few days a new sticky mat formed.

John Gibb, a bleacher in Paisley, Scotland, built a municipal water, slow sand filtration system in 1804. Gibb supplied his bleachery and the entire town with filtered water he delivered by cart. By 1807 his sand-filtered system was pumping and piping filtered water in Glasgow. Robert Thom, an engineer in Paisley, improved Gibb's filtration system with his invention and patenting in 1827 of a water backlash process that reversed the direction of the waterflow and much more effectively cleaned the sand filter. In Paris city officials in 1806 opened a large slow sand and charcoal filtration water treatment plant to purify water from the Seine River. The water settled for 12 hours before filtration. Every 6 hours workers renewed the sticky surface mat that filtered the water. The plant operated for 50 years.

James Simpson (1799–1869), an English scientist and civil engineer, initiated the construction of slow sand filtration systems to treat municipal water supplies throughout the United Kingdom. In 1829 he constructed London's first slow sand filtration system, a one-acre filtration system that the Chelsea Water Works Company used to purify water from the Thames River. Yet two decades elapsed before London in 1852 passed the Metropolitan Water Act that required filtration of its water supplies. Even then, Faraday at the Royal Institution in an 1855 letter to the *London Times* complained about the river's horrible smell and the dense feculence he observed.

The water still smelled in 1868 when Edward Frankland (1825–99), London's official water analyst from 1865 to 1876, carried out a four-month laboratory filtration study of London's raw sewage-laden water. Frankland passed the water through columns packed with substances such as coarse gravel and peaty soil. His study, which included chemical and bacterial analyses, left little doubt about the poor quality of the Thames water. Frankland issued monthly reports on London's water supply until his death in 1899.

Frankland's filtration study preceded a major scientific and medical breakthrough in water purification that occurred in Germany. In 1892 Koch first clearly demonstrated the effectiveness of slow sand filtration in reducing cholera deaths in his comparative study of the drinking water in two German cities, Hamburg which did not filter its water and Altona which did. Both of these contiguous cities drew their water from the Elbe River, but Altona, which had to filter its water, now badly polluted because of its location downstream from Hamburg, experienced far fewer deaths from cholera than Hamburg, 230 per 100,000 population versus 1,344 per 100,000 population.

In the United States, Richmond, Virginia, on the James River, operated the first slow sand filtration plant. Richmond began filtering its water in 1832 and had a centralized treatment facility serving 295 customers in 1833. Other cities with slow sand filtration systems included Elizabeth, New Jersey, in 1855; Poughkeepsie, New York, in 1872; Lawrence, Massachusetts, in September 1893; Philadelphia in 1902–11; Pittsburgh in 1908; and Evanston, Illinois, in 1914. Toronto, Ontario, Canada, had a slow sand filtration plant that filtered 180 million liters per day in 1909–11. The filtration increased to 275 million liters per day in 1914–17.

The slow sand filter installed at the Lawrence plant cost $31,000 per acre of filtering surface, whereas the 8 filters (filter beds) installed at the Albany, New York, plant in 1899 cost $45,600 per acre of filtering surface. The total cost of the Albany water treatment plant with a daily capacity of 60 million liters (15 million gallons) was $499,890. The five filters at the Philadelphia plants each cost $145,000 and $187,000 per acre including the cost of the 66 million liter (16.5 million gallons) reservoir. The cost of the largest plant, at Torresdale (1909), with a daily capacity of 960 million liters (240 million gallons), totaled $9,208,000. At the Pittsburgh plant each of the 46 one-acre covered filters cost $78,000, the complete water treatment plant of 480 million liters (120 million gallons) daily capacity cost $5,970,971.

TYPHOID FEVER DEATH RATE PER 100,000 POPULATION			
City	*Several Years Before Installation of SSF*	*Several Years After Installation of SSF*	*Decrease Percentage*
Cincinnati	54	10	81
Columbus	83	16	80
Louisville	58	17	71
Lawrence (MA) operational 1893	122	17	86
Minneapolis	35	4	88
New Orleans	40	23	43
Philadelphia operational 1906	63	20	68
Pittsburgh operational 1908	120	18	85
Hamburg (Germany)	51	9	87

The circumstances that led to construction of the slow sand filtration plant in Lawrence were similar to the events that precipitated Koch's 1892 study. The Massachusetts State Board of Health in 1887 had established an experiment station in Lawrence to study the treatment of drinking water obtained from the sewage-polluted Merrimac River. In the midst of the water tests the 1890–91 typhoid fever epidemic struck Lawrence, an industrial city of 45,000 with a 45 percent immigrant population.

Hiram Mills (1836–1921), chief engineer, and Allen Hazen (1869–1930), chief chemist at the experiment station, in 1893 designed and constructed a slow sand filtration system to remove typhoid fever bacteria, and in post-1893 studies the biologist William Sedgwick (1855–1921) at MIT established its effectiveness in removing bacteria. The number of bacteria fell from 12,700 per cubic centimeter of water to 70 per cubic centimeter of water, Lawrence's typhoid fever deaths decreased by 86 percent following the installation of the slow sand filtration plant. Typhoid fever deaths in the United States dropped from 48 per 100,000 population in 1890 to 13 per 100,000 population in 1917 because of filtration and the introduction of chlorine early in the twentieth century.

Slow sand filtration remained widely used in the United States until about 1920. It proved uneconomical for large-volume water purification systems because slow sand filters require large land areas to supply sufficient amounts of filtered water. Average cost in 1917 was about $60,000 per acre of filtering surface, daily capacity per acre was 12 million liters (3 million gallons). It has limited application in the United States, mainly for small-scale water purification systems, but remains in use in Europe and Asia.

LAWRENCE, MASSACHUSETTS TYPHOID FEVER DEATHS	
Years	*Deaths/100,000*
Years before SSF operational 1888–92	129
Year SSF operational 1893	60
Years after SSF operational 1894–98	17

Rapid sand filtration first appeared in the mid-1880s and gradually replaced many of the older slow sand filtration systems. Rapid sand filtration systems generally cost less to construct but more, sometimes twice as much, to operate and maintain as slow sand filtration systems. In rapid sand filtration water passes through a coarse sand filter, having larger pore spaces (0.55 millimeters and bigger) between the sand grains, at the rate of 3,200–4,800 liters per hour per square meter (2–3 gallons per minute, or 120–180 gallons per hour per square foot) of sand area. For the same area, rapid sand filtration acts 20 to 40 times faster than slow sand filtration

The introduction of coagulation hastened the development of rapid sand filtration. Coagulants are inexpensive chemical compounds such as calcium hydroxide also called milk of lime $Ca(OH)_2$, iron (III) chloride $FeCl_3$, iron (II) sulfate $FeSO_4$, and alum $K_2SO_4 \cdot Al_2(SO_4)_3 \cdot 24H_2O$, a dye fixer. Alum cost $20 per ton in 1916, iron (II) sulfate $11 per ton. When added to turbid water, they caused undissolved solids in the water to clump together (flocculation) because of electrostatic attraction. The clumping increased particle size and weight, enough to make the solids sink to the bottom of the container.

Because of its rapidity, rapid sand filtration with coagulants added removes some bacteria, protozoa, and viruses, but it is not as effective as slow sand filtration in removing bacteria and protozoa and removes few viruses.

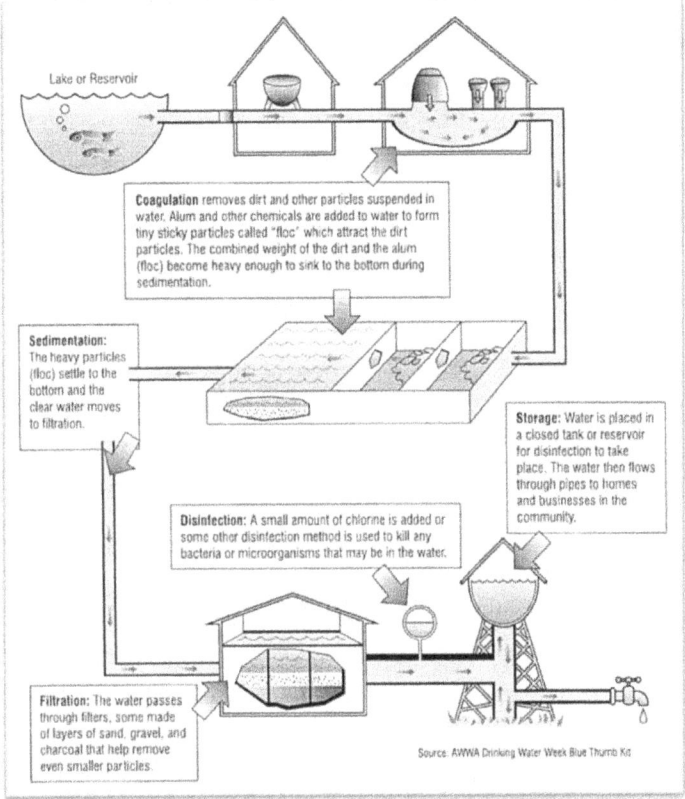

Courtesy of the EPA.

Figure 31 Water Distribution and Purification System. Follow a drop of water from the source through the treatment process which may differ in communities depending on the quality of the water entering the plant. Groundwater is located underground and typically requires less treatment than water from lakes, rivers, and streams.

Isaiah Smith Hyatt (1830–85) in Morristown, New Jersey, an associate of New Jersey's Newark Filtering Company and older brother of John Wesley Hyatt (1837–1920) celluloid's inventor, patented a coagulation-rapid filtration process in 1884. He used iron (III) sulfate $Fe_2(SO4)_3$, iron (III) chloride $FeCl_3$, potassium permanganate $KMnO_4$, or other coagulating agent (a positively-charged chemical coagulant), to attract negatively-charged impurities in water (US Patent 293, 740, 19 February 1884). The coagulation or curdling produced a jelly-like substance, or flocculent (floc), that filtration easily removed. The Newark Company in 1885 installed the first Hyatt alum filter and coagulant system in Somerville, New Jersey.

By the end of the nineteenth century, coagulation-rapid sand filtration systems outnumbered slow sand systems 10 to 1. Worcester, Massachusetts, in 1890, Louisville, Kentucky, in 1896, and New Milford, New Jersey, in 1903, were the first cities to apply coagulation and filtration. Worcester's chemical precipitation plant had six chemical precipitation settling basins, each 662 feet × 100 feet × 7 feet, for screening and treating raw sewage with calcium hydroxide. Each settling basin had the capacity to treat 350,000 gallons of sewage daily, although the plant's design enabled it to treat 3 million gallons of sewage daily. The treated sewage passed through a mixing channel into the six settling basins connected in series in about six hours. After two hours, when the top water layer became calm, workers drew it off for filtration and passed the bottom sludge to a centrifugal pump that discharged the sludge into nearby lagoons. The addition of ten more settling basins in 1893 increased the plant's capacity to 5.5 million gallons and the total cost of the precipitation plant to $200,000.

The combined coagulation-rapid sand filtration process passed its first crucial test in 1908. In 1884–85, the Louisville Water Company's chief engineer, Charles A. Hermany (1830–1908) who in 1861 succeeded Scowden now in Louisville, had tried settling the turbid water of the Ohio River in a reservoir but failed mainly because the reservoir was too small to provide clear water. George W. Fuller (1868–1934), a MIT-educated chemist whom Hermany and Louisville's Board of Water Works hired in 1895, provided the solution. He pretreated the water with a combination of coagulation with alum and sedimentation before filtration. Fuller published a report of his successful combined treatment in 1898. Ten years later, in 1908, Louisville's Crescent Hill Water Filtration Plant tested its first filters and in 1909 began operation. Louisville's combined treatment eliminated turbidity and claimed 99 percent removal of bacteria from its water.

The success of Louisville's coagulation and sedimentation treatment contributed to Fuller's reputation as the father of American sanitary engineering. With the introduction of chlorine in 1914 Louisville's death rate from typhoid fever dropped 80 percent from 71 per 100,000 in 1907 to 14 per 100,000 in 1915. Its coagulation and sedimentation treatment became the model and standard and paved the way for widespread adoption of rapid sand filtration in the United States.

Chlorination

The final advance in the water purification process was the addition of chlorine gas or hypochlorite compounds to kill deadly bacteria and disinfect water. The

Swedish chemist Karl Scheele (1742–86) at Uppsala discovered chlorine gas in 1774 when he reacted hydrochloric acid with manganese dioxide. Establishing the chemical composition of substances posed a serious problem in Scheele's time, and decades passed before Humphry Davy (1778–1829) at the Royal Institution in 1810 identified the gas as an element and named it chlorine, derived from the Greek *chloros* for pale green.

Chlorine's potential as a commercial product began in 1785 when Claude Louis Berthollet (1748–1822) in Paris first observed the bleaching properties of chlorine solutions on cloth. A few years later, in 1789, chlorine found limited use as a textile bleachery solution, named *Javelle water*, for the French town in which textile producers prepared the bleach by passing chlorine gas into an aqueous solution of potassium carbonate (potash). The reaction gave a potassium hypochlorite (KClO) solution. Using a sodium carbonate (soda ash) solution gave an equally effective sodium hypochlorite (NaClO) bleaching solution. Both hypochlorites are liquid compounds in their normal state.

Another 40 years elapsed before chlorine solution found use as a germicide and disinfectant. Antoine Labarraque (1777–1850), a Paris chemist, in 1820 experimented with sodium chlorite solution as a disinfectant, and in 1827 Thomas Alcock (1784–1833), a London surgeon, published an essay in which he recommended using solutions of sodium hypochlorite or calcium hypochlorite, $Ca(ClO)_2$, to disinfect hospitals, reservoirs, and sewers.

The major breakthrough that supported Labarraque's and Alcock's research and led eventually to the widespread application of chlorine solution as a disinfectant occurred in 1847. The Hungarian physician Ignaz Semmelweis (1818–65) at the General Hospital in Vienna ordered his medical students to wash their hands with bleaching powder (chlorinated lime or chloride of lime) solution, $CaCl(ClO).4H_2O$, each time they treated a new patient.

Semmelweis suspected the students were spreading deadly germs when they went directly from performing autopsies and dissecting corpses to the maternity ward and other wards for clinical instruction without washing their hands. The students examined pregnant women or attended to women in labor and, as Semmelweis proved, their germ-bearing hands caused the hospital's high female mortality rate from puerperal fever (child bed fever). The medical community largely opposed or ignored Semmelweis' results which showed that the female mortality rate at the General Hospital dropped significantly from 18 to 2.4

percent. Physicians argued that they saved lives and did not contribute to their patients' deaths.

Snow's addition of chlorine to disinfect a pump water supply in London following the 1854 cholera outbreak, likely the first known application of chlorine gas to disinfect water, supported the earlier evidence demonstrating the effectiveness of chlorine and hypochlorites as disinfectants. Koch's 1881 demonstration of the deadly effect of calcium hyprochlorite on bacterial cultures provided more evidence. Yet chlorination for water purification began to gain general acceptance first in Europe and then in the United States only near the turn of the century. Perhaps the public feared the consequences of drinking a liquid containing the deadly gas chlorine or the irritant and toxic bleaching agents, the hypochlorite compounds. At that time public waterworks added either chlorine gas or the equally effective hypochlorites or chlorinated lime to purify their water supplies. These compounds were safer, provided the same protection against pathogens as chlorine gas, but more expensive.

One of the earliest commercial-scale applications of chlorine added to water as a disinfectant occurred in 1893, the year after Koch's cholera study, when the city of Hamburg added chlorinated lime to its water. A second application followed in 1897 in Maidstone, Kent, England, when German Sims Woodhead (1855–1921), a Cambridge pathologist, used a sodium hypochlorite bleach solution as a temporary measure to sterilize potable water following a typhoid outbreak. The first municipal use of chlorine gas occurred in 1903 when the Belgium Ministry of Public Works added chlorine gas to disinfect drinking water in Middlekerke, Belgium. The next year, in 1904, London added sodium hypochlorite (10 percent available chlorine regulated to give a concentration of one part per million) to disinfect its drinking water to stop a typhoid fever epidemic that struck the city. Alexander Houston (1865–1933), a medical doctor and a director of the London Metropolitan Water Board, and his associate Cecil McGowan carried out the first systematic use of chlorine gas or a chlorine compound to treat a large water supply, namely London's.

Chlorine treatment began in the United States in 1908. The Jersey City Water Supply Company in Boonton, New Jersey, upon the recommendation of John Leal, a medical doctor and adviser to the Jersey City Water Supply Company, on 26 September 1908 added chlorine as sodium hypochlorite to its urban water supply, the Boonton Reservoir. The reservoir located 23 miles from Jersey City, a city of 200,000, began providing Jersey City with water in 1904. A Rockaway

River tributary was polluting the reservoir, and Leal, working with Fuller of Hering and Fuller, Hydraulic Engineers and Sanitary Experts, designed the chlorine purification system.

Fuller and his partner Hering, father of the modern American municipal sewerage system, had excellent reputations in sanitary engineering. Their company established in New York City in 1901 manufactured water treatment equipment. Jersey City adopted their chlorination system because adding the cheap and abundant hypochlorite compound to the reservoir spared it the cost of having to build a more expensive sand filtration system. The chlorination cost 14¢ per million gallons of water, or $5.60 per day for Jersey City's 40 million gallon per day water consumption.

Only two days earlier on 24 September 1908, George A. Johnson (1874–1934) of Hering and Fuller installed a calcium hypochlorite chlorination unit at the Union Stock Yards' five million gallon per day Bubbly Creek Filter Plant in Chicago. Chicago's meatpackers, particularly Union Stock Yards, dumped blood, entrails, and other animal wastes into Bubbly Creek, the nickname of the badly polluted south fork of the Chicago River's south branch. The water also reeked of hydrogen sulfide gas and methane formed from the decomposing waste dumped in the water, hence its nickname, Bubbly Creek. Bubbly Creek provided the 475-acre Union Stock Yards' livestock with filtered drinking water treated with copper (II) sulfate as a disinfectant. But Johnson's investigation showed that the filtration and disinfection failed to reduce the water's high bacterial count and that the high bacterial count prevented the livestock from gaining weight. Underweight livestock seriously threatened the meatpackers' profits.

The chlorination unit that Hering and Fuller installed reduced the bacterial count from 225,000–1,390,000 bacteria per cubic centimeter of water to 1–55 bacteria per cubic centimeter, thereby purifying the Union Stock Yards' livestock water supply and ending the livestock's inability to gain weight. Their chlorination unit provided the first commercial-size chlorination treatment in the United States, whereas Jersey City provided the United States' first urban chlorination treatment. Other cities that introduced chlorination included Albany, New York, 1909; Minneapolis in 1910, Cleveland and Pittsburgh in 1911, Louisville in 1914, Salt Lake City in 1915, Redding, California, in 1916, and Evanston Illinois, in 1921. By 1918 over 1,000 American cities were treating three billion gallons of water per day.

Already in the 1890s American industries were producing the chlorine-containing chemical compounds so crucial to water purification. The Mathieson Alkali Works, which in 1892 constructed its Saltville, Virginia, plant, began producing bleaching powder (chlorinated lime) obtained from the electrolysis of brine in 1896. In 1897 Mathieson was producing chlorine-alkali products at its plant in Niagara Falls, New York. The Dow Chemical Company in Midland, Michigan, also began manufacturing bleaching powder in the 1890s. Later in 1924, James Cloyd Downs (1885–1957), a chemist at the Roessler and Hasslacher Chemical Company in Niagara Falls, New York, patented a widely used electrolytic process for the production of chlorine from molten sodium chloride. DuPont purchased Roessler and Hasslacher in 1930. The Solvay Process Company in Hopewell, Virginia, in 1936 introduced an alternative chlorine production process called the nitrosyl process.

Apparatus to deliver measured amounts of chlorine followed in 1913 when two unemployed engineers Charles Wallace (1885–1964) and Martin Tiernan (1883–1968) in New York City and later in Belleville, New Jersey, invented the first practical and effective apparatus for the controlled feeding of chlorine gas. Their $150 chlorinator that six dry cells (6 × 1.5 volts) operated piped the equivalent of a thimble-full (2-7 ml) of chlorine gas into 1 million parts of water (1 part per million = 1 milligram per liter, 1 ppm = 1mg/L). The Boonton Reservoir that provided Jersey City's drinking water installed the first Wallace-Tiernan chlorinator that same year.

Unquestionably the most significant advance in chlorination came in 1919 just after World War I ended. Abel Wolman (1892–1989), an engineer at the State of Maryland Department of Health in Baltimore who later taught at Johns Hopkins, and his coworker, Linn Enslow (1891–1957), a Health Department chemist, developed a safe and reliable method for determining the appropriate dose of chlorine for purification of any water source. Their work advanced the 1908 kinetic studies of Harriette Chick (1875–1977), a research student at the Lister Institute of Preventive Medicine in London, and Herbert E. Watson (1886–1980) at University College, London. Chick developed the first chemical rate of reaction equation that related a disinfectant's effectiveness in killing an organism and its contact time with the organism. That same year Watson modified Chick's equation to include a coefficient of specific lethality to give a rate equation known as the Chick-Watson equation.

Wolman and Enslow noted that upon adding chlorine to water the amount of chlorine absorbed varied significantly depending on the water quality, and they introduced the idea of chlorine demand to indicate the rates of chlorine absorption. The chlorine demand represented the difference between the amount of chlorine added and the amount remaining after a specified time interval. The higher the chlorine demand, the higher the bacterial count and presence of other organisms, and the more polluted the water. Over an eighteen-month period in 1917–18, they tested 350 water samples from the Potomac River and samples from other rivers, measuring the bacteria, acidity, and factors that possibly affected the taste, turbidity, and purity of a specific water source. They compared the relative chlorine demands (the amount of chlorine absorbed) of water sources containing different amounts of inorganic, organic, or biological matter, and determined the chlorine dose for safe water treatment.

Wolman and Enslow's initial chlorine concentration was 1 mg/L (1 ppm), high enough to disinfect most drinking water sources, the water temperature 20°C (68°F), and the time intervals ranged from 5 to 60 minutes. They found that a five minute time interval at 20°C provided reliable results, with very little gain resulting from a 30 or 60 minute interval. They recommended adding another 0.2 mg/L of chlorine as a safety factor but opposed maintaining residual chlorine in the water distribution system. The residual chlorine, they said, would result in possible taste and odor problems. The Wolman-Enslow chlorine demand method remains the standard water pollution testing treatment used globally.

Two other important water purification advances occurred during this period. Twenty-two waterworks supervisors from Illinois, Indiana, Iowa, Kansas, Kentucky, and Tennessee met in 1881 at Washington University in St. Louis, Missouri, to discuss management of waterworks. At the meeting they founded the American Waterworks Association. They aimed to improve the quality and supply of water. The second advance occurred when the Department of Treasury enacted the first standards required for disinfecting drinking water. It set a maximum of two coliforms (two detected pathogens) per 100 milliliters (mL) in 1914, lowered the standard to one coliform per 100 mL in 1925 and included standards for copper, lead, and zinc. The standard applied only to interstate water carriers, such as trains, buses, and boats and to contaminants capable of causing contagious diseases. The 2014 standard for total coliforms is zero.

As a result of the improvements in water purification that began in the late nineteenth and early twentieth centuries, the number of deaths in the United

States from typhoid fever, the most virulent of water-borne diseases, dropped from 27,000 in 1900 (36 per 100,000 population) to one-eighth or about 3,400 (3 per 100,000) in 1930. During this time the population increased from 75.9 million to 120.7 million. Deaths in Chicago from typhoid fever from 1860 to 1900 had averaged 65 per 100,000 population per year. In Pittsburgh in 1907, typhoid fever killed 648 of its 535,330 (121 per 100,000) inhabitants. In Cleveland the death rate from typhoid fever dropped from 110 people per 100,000 in 1900 to one per 100,000 in 1930. Cases of cholera and dysentery also decreased significantly.

Later revisions of the drinking water standards occurred in 1946, 1962, 1974, 1986, and 1996. The 1974 Safe Drinking Water Act for the first time required all public water systems to comply with the government's health standards, exempting only non-community water systems. In the United States the Environmental Protection Agency's (EPA, established in December 1970) 2013 minimum-maximum chlorine dosage for drinking water is 0.2–4.0 mg/L. The World Health Organization (WHO) has recommended a maximum chlorine concentration of five mg/L.

III. INDUSTRIAL WATER POLLUTION

A second water pollution problem that emerged at this time resulted from the construction of industrial plants on the banks of rivers and lakes to take advantage of cheaper water transportation. Plants, such as the Carnegie and US Steel plants in Pennsylvania on the banks of the Monongahela River used the river to dump industrial waste. A similar problem arose in the Niagara River-Great Lakes region of New York State. Chemical companies such as DuPont, Carborundum, Mathieson (in 1954 Olin Mathieson), and Hooker Electrochemical Company (acquired by Occidental Petroleum in 1968), had located there in the late nineteenth and early twentieth centuries to take advantage of the recently developed hydroelectric power generating stations at Niagara Falls.

Carborundum, a company that Edward Acheson (1856–1931) established in Monongahela City, Pennsylvania, in 1894 and relocated to Niagara Falls in 1895, constructed a plant with electric furnaces to manufacture carborundum (SiC), an abrasive crystalline material that Acheson invented in 1891 (patented in 1893) and artificial graphite in 1893 (patented in 1896). Hooker, established in 1903 as the Development and Funding Company, began electrolyzing brine for the production of chlorine at its Niagara Falls plant in 1904. Chlorine soon found an important use as the main disinfectant added in water purification processes, and in the production of sodium hydroxide.

The increased number of ships on the lakes and rivers, transporting manufactured products, raw materials, grains, and other agricultural produce contributed further to water pollution by discharging their waste into these waterways.

IV. AIR POLLUTION

Air pollution has a long history, dating from the first natural disasters such as volcano eruptions, the burning of wood, and roasting (smelting) of copper, lead, iron, aluminum, and other metallic ores, but it became a serious problem beginning with the industrial revolution in the late 1700s–early 1800s. The construction of the first industrial plants in England and their consumption of large amounts of coal for fuel, followed by a similar pattern of industrialization in Europe and the United States, brought considerable detrimental change to the environment. The rapid pace of industrialization in the last half of the nineteenth century and the emergence of the automobile industry in the first decade of the twentieth century added to the growing problem of air pollution.

The automobile put an end to the bodies of dead horses on city streets and to the tons of manure that horses dropped on the streets. New York City in 1880 removed 15,000 dead horses from its streets, Chicago removed 9,202 in 1916. Estimates indicate that about three million horses provided valuable services in American cities in the late 1800s–early 1900s and that a healthy horse deposited about 20 pounds of manure daily. If not removed before it dried, which seldom occurred, the manure became flaky and easily dispersed by the wind, thereby increasing the spread of disease-carrying manure. Automobile exhaust, however, put large amounts of nitrogen, water vapor, carbon dioxide, and smaller amounts of nitrogen oxides, carbon monoxide, hydrocarbon vapors, and sulfur dioxide (traces) into the atmosphere. The nitrogen oxides have contributed to ozone production and photochemical smog.

Industrial plants and petroleum refineries also released tremendous quantities of noxious gases (sulfur dioxide, nitrogen oxides), hydrocarbon vapors, carbon dioxide, and particles of soot, smoke, and dust. In urban areas sulfur dioxide concentrations increased steadily during the second half of the nineteenth century, not peaking until 1920–50. Eastern seaboard cities continued to use residual fuel oil, with improvements in sulfur air quality not occurring until the

1960s. As a result, New York City, Chicago, and other large cities have experienced acid rain since the 1930s when air quality studies began, and most northwestern, medium-sized and smaller cities have experienced acid rain since 1950.

Nitrogen oxides emissions from coal-fired power plants in the northeast also increased steadily in the nineteenth century and contributed to the acid rain problem. The rise continued into the twentieth century augmented significantly by automobile exhaust emissions so that by 1970 the nitrogen oxides concentrations were three times that of 1920.

American Cyanamid Company, established in 1907, constructed its first plant in 1909 near Niagara Falls, Ontario, and since the early twentieth century the plant, which had a capacity of 4,500 tons cyanamide per year, has discharged clearly visible reddish-brown or orange nitrogen dioxide fumes into the atmosphere. DuPont, Hooker, and other chemical plants in the Niagara Falls, New York area also discharged large quantities of polluting vapors into the atmosphere, and the petroleum refineries in Texas, Oklahoma, and Louisiana have had a similar disposal problem with hydrocarbon and sulfur dioxide emissions.

Acid rain and smoke

Most of the nineteenth-century publications on industrialization's disastrous effects on air quality dealt with smoke pollution from burning soft bituminous coal. Carbon particles (soot) resulting from the combustion of coal and suspended in air constituted coal smoke commonly called smoke. Fog is a suspension of water droplets in air. When smoke combines with fog, the combination of smoke and fog results in smog. The combustion of coal usually produces sulfur dioxide and nitric oxide, and their release and presence in the atmosphere accounts for the acidity of fog, smog, and rain. Few people recognized the problem of acid rain. The first and possibly most significant analysis appeared in an article that the Scottish chemist Robert Angus Smith (1817–84) published in 1852.

Smith, whom Queen Victoria (1819–1901) in 1863 appointed the first inspector in the Alkali Acts Administration, accidentally observed an unusual acidity in Manchester's rains during the winter of 1852. He believed the soot particles released into the atmosphere from burning bituminous soft coal, and the cause of the city's smog, were also the reason for the acidity of the precipitated fog. His 1852 article "On the Air of Towns" reported cases of stunted tree growth and an acidity of the air that was at times high enough to turn litmus paper red immediately during a rain. The acidity resulted from the conversion of sulfur in the coal to sulfur dioxide gas, and its absorption by soot. Smith also estimated the amounts of carbonic acid and organic matter in the air and compiled figures showing a higher death rate in urban areas because of atmospheric pollution.

In a final set of experiments in which he bubbled polluted air through blood samples of inhabitants from different towns including Manchester, Smith concluded that the town's atmosphere had a peculiar effect on the blood's state. It caused a "remarkable reddening" "distinctly perceptible to the ordinary eye." Smith coined the phrase *acid rain* in his 1872 book *Air and Rain*.

Despite the significance of Smith's discovery of rain's acidity, his work failed to make an impact in its day. The explanations below answer the question: Why not? In the opening lines of the preface to *Air and Rain* Smith wrote, "This volume could not have been made interesting to readers without sacrificing that which I suppose to be its best point—namely the large number of facts. Another more popular in character might perhaps be made to follow with propriety. We have been wearied with speculations about the Air, and it seemed to me advisable to begin building a systematic foundation with experiments which may be repeated frequently, or they have already been repeated."

Smith consciously had chosen not to write *Air and Rain* as a popular account but as a compilation of the experimental evidence he had accumulated for years to highlight the damaging effects of pollution to air, land, and water. Sir Philip Joseph Hartog (1864–1947), an English chemist and educator in London who early in his career taught briefly at Owens College at Manchester and wrote science articles for the *Manchester Guardian,* made this point in the sketch he published on Smith in 1897–98 in the *Dictionary of National Biography*.

Hartog added a second reason for Smith's neglect and his obscurity. He believed that the attention scientists paid to Pasteur's work on the germ theory diverted attention from the inorganic impurities in air, which included the acids, and hence from the significance of Smith's work. The English physicist, chemist, and historian of science, Sir Thomas Thorpe (1845–1925), who graduated from Owens College and probably knew Smith, arrived at the same conclusion writing that "as the chemist of sanitary science, Smith worked alone." Smith never married and left no direct heirs. His nephew, W. Anderson Smith who in 1873 migrated to Melbourne, Australia, established a scholarship in sanitary science at Manchester in 1928 in honor of his uncle.

What Hartog, Thorpe, and other scientists of that time knew but left unsaid was that Smith had called attention to the other side of industrial development, the obvious pollution not only of the atmosphere but of land and water. The nineteenth-century was a time of tremendous industrial development in the western world. Industrial development meant progress and progress meant a superior culture. The scientific community, with few exceptions, and most of society, accepted the destructive effects of pollution as the price of progress.

We hear the same false argument about pollution and job loss. We can't reduce industrial pollution because the expense of doing so would result in plant closings and loss of jobs. Those who questioned industrial progress were powerless to do anything anyway. Whatever and whenever protest emerged, the protests were usually and justifiably against the laborers' inhumane and brutal working and living conditions as described in several well-known nineteenth-century books, such as Charles Dickens' *Hard Times* (1854) and Henry George's *Progress and Poverty* (1879). Protest about the environment seldom occurred.

Another chemist, Lajos Ilosvay (1851–1936) at the Budapest Technical University, in an 1889 article on nitrous acid in the atmosphere, established the presence of nitrogen compounds (nitrates) on the surface of grass, leaves, and branches and the presence of nitrous and nitric acid in the water of different kinds of soil. Ilosvay's article published in the *Bulletin de la Société Chimique de Paris* showed that the same soils in the presence of pure air contained no nitrogen compounds even after 10 days exposure. After exposing the soils to ordinary air for 26 hours he easily detected the presence of nitrous acid. Ilosvay did not discuss the formation of acid rain and its consequences, but his work clearly substantiated earlier work on acid rain.

In an important 1895 study on air quality and the acidity of the atmosphere because of coal burning, the chemist Charles F. Mabery (1850–1927) at the Case School of Applied Science in Cleveland detected the presence of excessive amounts of sulfurous-sulfuric acids and nitrous-nitric acids in Cleveland's air. Although the decay of plant and animal bodies released ammonia and ultimately nitric acid, the main source of these acids was the quantity of coal consumed in

Cleveland. It totaled 117,157 tons of clean burning (smokeless) anthracite and 924,602 tons of dirtier bituminous coal in 1889. Bituminous coal, according to Mabery, contained on average one percent sulfur, which although low, nevertheless released 9,246 tons of sulfur during combustion, or the equivalent of 80 tons daily or 28,305 tons of sulfuric acid annually.

The quantity of sulfur released varied throughout the city, but samples from one area totaled 11,851 grams (26 pounds, 1 pound = 454 grams) of sulfuric acid per 1,000,000 cubic meters of air and 12,512.6 grams of sulfurous acid per 1,000,000 cubic meters of air. An atmospheric volume of 1,000,000 cubic meters, a commonly-used volume, corresponded to the volume of air contained in an area of one square mile and 1.25 feet deep.

Mabery compared these figures to studies done in London and Manchester, England, and in Lille, France, which also had high concentrations of sulfuric acid because of the large quantities of smoke released during the combustion of bituminous coal. In London, the weight of sulfuric acid present in 1,000,000 cubic meters of air was 1,693.94 grams and for Manchester 2,443.74 grams. For Lille, the figures were two cubic centimeters of sulfurous acid per liter of air and 0.022 grams of sulfuric acid per liter of rain water or 2,200 grams per 1,000,000 cubic meters (1 liter = 10 cubic meters).

Mabery attributed the recent increase in the sooty condition of Cleveland's atmosphere to rapid growth in wealth and population which resulted because of Cleveland's proximity to immense deposits of cheap fuel. The consequences, he said, were clearly detrimental. In the atmosphere of a sooty city, as everyone knows, the nails of a roof are nearly consumed after 10 or 15 years, whereas nails lasted 75 years in the country. Soot, which readily absorbed sulfuric acid, when deposited on trees, contributed to their decay and destruction. The soot that landed on the chemical laboratory's window sill when washed with water and tested with litmus paper gave a strong acid reaction and when analyzed showed the presence of sulfuric acid.

The devastating effect of sulfuric acid was most apparent in libraries. Mabery noted that in 1879 his colleague William Ripley Nichols (1847–86) at MIT had demonstrated the absorption of sulfuric acid in book bindings, causing the bindings to crumble after a time. Mabery concluded with the comment that a sooty atmosphere is not conducive to either personal comfort or to health.

Most of the articles published on air pollution at that time dealt with the passage of city smoke ordinances, such as those passed in Chicago in 1881 and in St. Louis in 1892. Both situations stand out because despite the ordinance, decades elapsed before either city took any effective action. In Chicago 56 years passed before city officials realized that their marketing, transportation, and fuel facilities made inevitable Chicago's expansion into an industrial center and that the time to study the control of sulfur dioxide had now arrived. Unfortunately, the city had neither records of sulfur dioxide's harm to health, vegetation, and building materials, nor methods of abating sulfur dioxide.

The lack of action in St. Louis easily matched Chicago's. St. Louis had become aware of its smoke pollution problem in 1864 as a result of a court case between two neighboring property owners. However, nearly 30 years of anti-smoke protests followed before the city in 1892 issued its first engineering report on smoke and passed its first smoke ordinance. Another 34 years lapsed before St. Louis established a Citizens Smoke Abatement League in 1926 and another 11 years before it adopted its first effective legislation in 1937.

The 1937 ordinance required washing high ash, high sulfur Illinois coal, crushed to less than two inches in size. This consisted of feeding the crushed coal into a large water-filled tank, causing the coal to float to the surface and the sulfur impurities to sink to the tank's bottom. The washing removes only sulfur that is chemically combined with iron atoms and present in coal as iron pyrite, or fool's gold (FeS_2). Washing does not remove sulfur that is chemically combined with the carbon atoms that make up the coal.

The coal owners contested the washing in federal court, but the court ruled that this clause came within the police powers of a city attempting to regulate a nuisance and allowed the ordinance to stand. Washing greatly reduced the coal's sulfur content, and once the washing became effective, St. Louis experienced its first major improvement in air quality. A 1940 amendment required either hand-firing with smokeless fuel or the mechanical-firing of all city fuel-burning plants. Mechanical firing was to ensure a more complete combustion by providing a sufficient supply of air thereby preventing a too-rapid and incomplete combustion, maintaining a high-enough furnace temperature, and facilitating a furnace design of proper size, form, and spatial proportions.

In the winter of 1940–41 St. Louis's inhabitants burned nearly 1,250,000 tons of smokeless fuel. When the ordinance first passed in 1937, only about 250,000 tons were available in the city's coal market. The results that first winter exceeded the hopes of the most optimistic. Whereas, during the previous heating season, the United States Weather Bureau had recorded a total of 716 hours of thick and moderate smoke, the same period a year later produced only 197 hours, or a reduction of more than 72 percent.

By 1908, Buffalo, New York City, Rochester, and Syracuse, New York; Detroit; Washington, DC; Baltimore; Milwaukee; Minneapolis; Boston, and Springfield, Massachusetts; Toledo, Cleveland, Dayton, and Cincinnati (November 1881) Ohio; Pittsburgh; and Indianapolis, in addition to St. Louis and Chicago, had enacted smoke prevention legislation. In Indianapolis the Board of Trade in 1908 had to appoint a committee to determine the source of a dense black smoke that lingered over the city. To locate the polluters the state statistician sent out forms requiring factory and large building owners to report the coal types and grades they were consuming.

In Pittsburgh opponents of industrial smoke polluters in 1907 established the International Association for the Prevention of Smoke, the first major anti-air pollution organization, renamed the Air and Waste Management Association in 1990. Pittsburgh however, had one of the most lenient smoke ordinances of any large city in the United States in 1914. Of the stacks operating in its 152 stationary plants 40 percent were violating the city's smoke ordinance, as were 50 percent of the stacks on the city's front-fed stoker plants. By 1916, 75 American cities had anti-smoke ordinances, but violations were numerous and flagrant. Unfortunately in Pennsylvania and some other states such as West Virginia and Ohio, the sulfur content of the bituminous coal was often in the 3–10 percent range.

Nevertheless society began to pay attention to the economic loss from smoke pollution. In the next decades the article "The Dividends that Float up the Chimney," published in *Scientific American* (1912) and others in the *Proceedings of the Smoke Prevention Association of America* (1923), and *Science News Letter* (1941), made strong cases for the monetary losses. The *Proceeding*'s article pointed out that in 1923 Chicago spent $15 million on the water department, $75 million on the sanitary and health department, and $25,000 for smoke prevention. It estimated that correct operation of heating equipment could eliminate 80 percent of soot and smoke emissions. Society could not estimate the dollar cost of soot and smoke on health, it said, nor calculate the cost of being sick or half sick, all of which should stimulate action to abate smoke.

Cottrell and the electrostatic precipitator: the first industrial environmentalist

Despite the lack of action on the acid rain issue and limited action on smoke pollution, the first practical anti-pollution device that dealt with these and other air pollutants appeared early in the twentieth century. Frederick Cottrell (1877–1948), a chemist at The University of California, Berkeley, in 1907 invented and in 1908 patented the Cottrell electrostatic precipitator (US Patent 895,729 of 11 August 1908). His precipitator removed the particles that made up the fumes escaping from an industrial smokestack by charging negatively a wire suspended vertically within a smokestack with a direct current, high voltage source of at least 30,000 to 80,000 volts. This created an electrostatic field that ionized (charged electrically) the air molecules in the smokestack so that the fume particles in the smokestack acquired a negative electric charge on contact with the air molecules. Upon colliding with the positively-charged walls of the smokestack, the negatively-charged fume particles lost their charge, causing them to precipitate (settle) at the bottom of the smokestack, preventing their escape, and making possible their collection and recovery as valuable products. The precipitator removed 90–98 percent of the particles that made up the escaping gases and vapors.

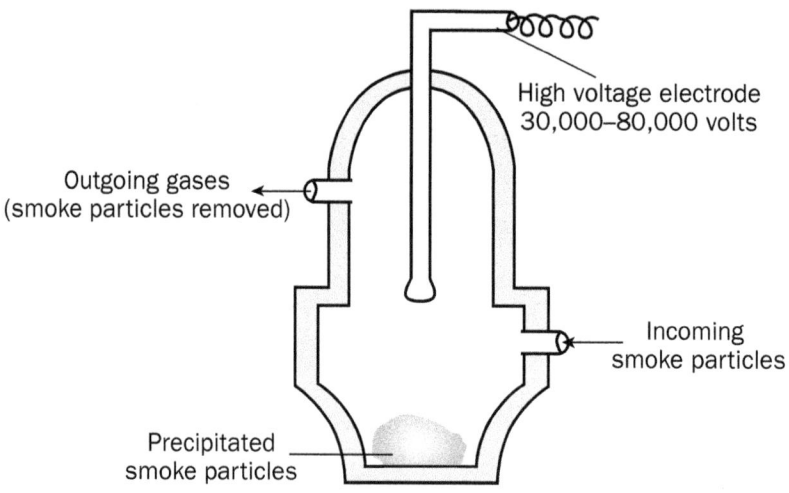

© Kendall Hunt Publishing Company.

Figure 32 Cottrell Electrostatic Precipitator. Light some smoke paper and drop it in the glass tube. When the tube is full of smoke, crank the Wimshurst machine to charge the electrode (wire) inside the tube. The smoke should diminish.

Cottrell first successfully demonstrated his precipitator in 1907 at DuPont's Pinole plant just north of Berkeley. The plant produced sulfuric acid and explosives and had a problem controlling acid mist formation and its escape into the air. A second successful demonstration followed in 1907 at the American Smelting and Refining Company in Selby, just north of San Francisco, where Cottrell's precipitator removed lead particles. Other successful applications occurred at Riverside Portland Cement Company in Crestmore, California, in 1911 and Anaconda Copper near Butte, Montana, in 1916 and 1919.

The precipitators, which cost as much as $100,000, also found use in California's oil industry by electrostatically removing water from oil by applying an electrostatic potential to the oil-water emulsion, thereby demulsifying it and reducing the water content in some cases from 50 percent to 1 percent. Electrostatic precipitators have continued to find wide use for dust collecting in cement plants, and in the removal of fly ash in coal-fired powerhouses, salt-cake fumes from black ash furnaces in paper mills, and acid mists from chemical plants. Other anti-pollution devices introduced after Cottrell's precipitator include centrifugal separators, packed beds, sulfur-absorbing limestone or lime scrubbers, and various types of airborne particle collectors.

Cottrell established the Research Foundation in New York City in 1912 to make research funds available to scientists for beneficial projects. The foundation provided funds for Robert Goddard's 1920s–30s research on rockets, Ernest O. Lawrence's 1930s research on the cyclotron, and for numerous other projects. The foundation continues to operate.

In the 1930s Cottrell invented the pebble-bed regenerative furnace to produce very high temperatures. It found limited use in the 1940s–50s in a high-temperature process to synthesize nitric acid from the nitrogen and oxygen present in the atmosphere. Cottrell hoped to provide a technologically-simpler alternative to the Haber high-pressure synthesis of ammonia from nitrogen and hydrogen, and ammonia's subsequent oxidation to nitric acid. The high-temperature process, however, was economically non-competitive with the high-pressure process.

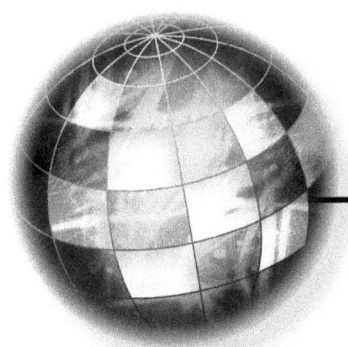

CHAPTER 5

CONCLUSION

In the first half of the nineteenth century, technology led science. This means that inventors, such as Cyrus McCormick (1809–84), Charles Goodyear, Samuel Colt (1814–62), and others, invented machines, devices, or processes that operated successfully, but they did not fully understand the scientific laws and principles that their inventions exemplified. Their inventive process was empirical, not theoretical, and included such factors as spatial thinking or visualization of their end product, serendipity, and even necessity. Technology still led science in the second half of the century, but science had made significant gains. The Gilded Age inventions that transformed the United States into the world's technological leader and an industrial giant required an increasing understanding of the laws and principles of science. This occurred in the major industries that emerged after the Civil War, in steel; the electrolytic production of aluminum; and in the petroleum, communications, and electrical industries. The increasing understanding of scientific laws and principles was crucial in transforming the United States from an agricultural to an industrial nation. The conversion from steam power to electrical power played a major role.

Twain called the last decades of the nineteenth century the Gilded Age, meaning that it dazzled on the surface but was corrupt or base beneath the surface. The analogy fits very well because the application of laissez-faire economics and Social Darwinism revealed the baseness and selfishness beneath the razzle and dazzle of technological triumph. Concern for the environment, despite the first recognition of acid rain and climate change (global warming), remained a cause for the future. Robert Angus Smith's work on acid rain went largely unnoticed until the mid-twentieth century.

George strongly criticized Malthus and Spencer, and Lloyd warned society that Social Darwinism brought out the worst in society not the best. Unfortunately, no one paid any attention to the wisdom of George, Lloyd, and others. Carnegie praised the know-it-all Spencer as the person to whom he owed the most, and he and Rockefeller ardently preached the Gospel of Wealth, telling everyone that their wealth made them special. They donated millions of dollars to various causes, most of which still bear their names, but millions of workers and their families bore an unnecessary burden that Carnegie, Rockefeller, and other Social Darwinists could have reduced had they only made an effort.

The table below, compiled from the analyses of historians such as Alfred Chandler, Thomas Cochrane, Carl Degler, and Robert Heilbroner, summarizes the technological transformation of Gilded Age America and its twentieth-century legacy.

SIX STAGES OF THE TECHNOLOGICAL TRANSFORMATION	
Period	*Event*
1815–50	The population moved westward.
1850–70	Railroads emerged.
1880–1900	Urban growth and the rise of cities provided markets for mass-produced goods.
1900–20	Electrical and automobile industries began to develop.
1920–50	Establishment and institutionalization of research laboratories, such as General Electric, Bell Laboratories, and DuPont Chemical occurred.
1950–2000s	Plastics, high tech, computer, and space industries emerged.

REVIEW QUESTIONS ON TECHNOLOGICAL TRANSFORMATION OF GILDED AGE AMERICA

1. What is the theme of this book?

2. Was the increase in the awarding of patents in the period 1860–1900 a good indicator of a technological transformation?

3. Define *laissez-faire* and Social Darwinism.

4. What factors contributed to the technological transformation that occurred in the United States in the period 1860–1900?

5. Were Europe and Asia undergoing a technological transformation during this same period?

6. What was the technological breakthrough or breakthroughs that led to the development of the steel, aluminum, petroleum, communications, and electrical industries in the period 1860–1900?

7. Who was the individual or individuals responsible for each of the breakthroughs discussed in question six? Were any the result of serendipity, the scientific method, or the deductive process?

8. Were any of the breakthroughs discussed in question six simultaneous or near-simultaneous inventions? Did any feuds result because of priority claims?

9. How does the open hearth steel process differ from the Kelly-Bessemer process?

10. Why did the steel industry emerge first in Pennsylvania and not elsewhere in the United States?

11. What were some of the other nineteenth-century advances in steel manufacturing?

12. Who were the organizers of the industries discussed in question six? What methods of organization did they invent or perfect?

13. What was the public's response to the technological transformation that occurred in the period 1850–1900?

14. Who were some of the well-known critics?

15. What is the warning given in Mark Twain's *A Connecticut Yankee in King Arthur's Court*?

16. What was the US Congress's response to the problems arising from industrial development? What action did Congress take? Why was Congress slow to act and when it did act were its actions effective?

17. What were the three major advances that occurred in water treatment in the late nineteenth, early twentieth century?

18. What was the main cause of nineteenth-century air pollution? Did anyone recognize acid rain as a problem?

19. Who is Frederick Cottrell and what did he invent?

20. Why should the public concern itself with air and water pollution?

21. Has concern for the environment improved in the twenty-first century?

22. Identify: Alexander Holley, Robert Mushet, spiegeleisen, Sidney Gilchrist Thomas, Charles Martin Hall, George Bissell, Titusville and Spindletop, Thomas Edison, George Westinghouse, Nikola Tesla, Emil Berliner, Alexander Graham Bell, Michael Pupin, Leo Baekeland, Henry George, Edward Bellamy, acid rain, Robert Angus Smith.

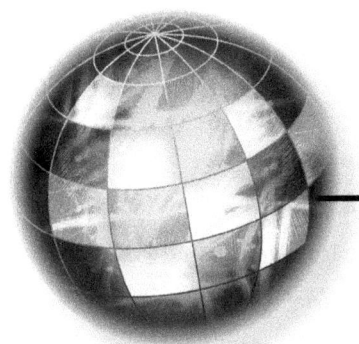

BIBLIOGRAPHY

Adair, Gene. *Thomas Alva Edison: Inventing the Electric Age*. New York: Oxford University Press, 1996.

Baker, Burton H. *The Gray Matter: The Forgotten Story of the Telephone*. (History of Technology, 26) St. Joseph, MI: Telepress, 2000.

Bellamy, Edward. *Looking Backward 2000–1887*. New York: Penguin Putnam, Signet Classic, 2000.

Berger, Michael L. *The Automobile in American History and Culture: A Reference Guide*. (American Popular Culture). Westport, CT: Greenwood Press, 2001.

Bessemer, Henry. *Sir Henry Bessemer, FRS: an autobiography; with a concluding chapter*. London: Offices of Engineering, 1905.

Black, Brian. *Nature and the Environment in Nineteenth-Century American Life*. Westport: Greenwood, 2006.

Boucher, John Newton and William Kelly. *A true history of the so-called Besscher process*. Greenburg, PA: Privately printed, 1924.

Brooks, John. *Telephone: The First Hundred Years*. New York: Harper & Row Publishers, 1976.

Burlingame, Roger. *March of the Iron Men: A Social History of Union Through Invention*. New York: C. Scribner's Sons; London: C. Scribner's Sons, ltd., 1938.

Burstall, Aubrey. *A History of Mechanical Engineering*. Cambridge, Mass.: The MIT Press, 1965.

Carnegie, Andrew. "Wealth," *North American Review*, vol. 148, issue 391, June 1889: 653–65.

Carr, Charles C. *ALCOA: An American Enterprise*. New York: Rinehart & Company, Inc., 1952.

Chandler, Alfred D., Jr. *Shaping the Industrial Century: The Remarkable Story of the Modern Chemical and Pharmaceutical Industries.* (Harvard Studies in Business History, 46). Cambridge: Harvard University Press. 2005.

Cheney, Margaret. *Tesla: Man Out of Time.* New York: Laurel, 1981.

Chernow, Ron. *Titan: The Life of John D. Rockefeller, Sr.* 2nd ed. New York: Vintage, 2004.

Cochrane, Thomas. *Frontiers of Change.* New York: Oxford University Press, 1981.

Cole, Jonathan R. "Two Cultures Revisited," *The Bridge*, National Academy of Engineering, vol. 26, no.3–4, Fall/Winter 1996: 16–21.

Collins, Theresa M., and Lisa Gitelman. *Thomas Edison and Modern America: A Brief History with Documents.* Boston: Bedford/St. Martin's, 2002.

Conot, Robert. *A Streak of Luck.* New York: Seaview Books; trade distribution by Simon and Schuster, 1979.

——. *Concise Dictionary of American Biography.* 3rd ed. New York: Charles Scribner's Sons, 1980.

——. *Concise Dictionary of Scientific Biography.* New York: Charles Scribner's Sons, 1981.

Davis, Devra. *When Smoke Ran Like Water: Tales of Environmental Deception and the Battle against Pollution.* New York: Basic Books, 2004.

Davis, Lawrence J. *Fleet Fire: Thomas Edison and the Pioneers of the Electric Revolution.* New York: Arcade Pub., 2003.

Degler, Carl. *Out of our Past*, 3rd ed. New York: Harper & Row Publishers, 1984.

Dubofsky, Melvyn. *Industrialism and the America Worker, 1865–1920.* New York: Crowell, 1975.

Essig, Mark. *Edison and the Electric Chair: A Story of Light and Death.* New York: Walker & Company, 2003.

Evans, Harold, Gail Buckland, and David Lefer. *They Made America: From the Steam Engine to the Search Engine: Two Centuries of Innovators.* New York: Little, Brown, 2004.

Evenson, Edward A. The Telephone Patent Conspiracy of 1876; the Elisha Gray-Alexander Bell controversy and its many players. North Carolina: McFarland & Company, Inc., Publishers, 2000.

Farber, David. *Sloan Rules: Alfred P. Sloan and the Triumph of General Motors.* Chicago: University of Chicago Press, 2002.

Foluché, Rayvon. *Black Inventors in the Age of Segregation: Granville T. Woods, Lewis H. Latimer, and Shelby J. Davidson.* Baltimore: The Johns Hopkins University Press, 2003.

Friedel, Robert, and Paul Israel. *Edison's Electric Light*. Baltimore: The Johns Hopkins University Press, 2010.

Furter, William F. *A Century of Chemical Engineering.* New York: Plenum Press, 1982.

George, Henry. *Progress and Poverty.* Garden City, New York: Doubleday, Page & Co., 1920. First publication date 1879.

Gibb, George S., and Evelyn H. Knowlton. *The Resurgent Years, 1911–1927: History of Standard Oil Company (New Jersey).* New York: Harper & Brothers, 1956.

Giberti, Bruno. *Designing the Centennial: A History of the 1876 International Exhibition in Philadelphia.* Lexington: University Press of Kentucky, 2002.

Gordon, Robert B. "The 'Kelly' Converter," *Technology and Culture*, vol. 33, no. 4, Oct. 1992.

Gray, Charlotte. *Reluctant Genius: Alexander Graham Bell and the Passion for Invention.* New York: Arcade Publishing, Distributed by Hatchette Book Group USA, 2006.

Gregor, Arthur. *Bell Laboratories: inside the world's largest communications center.* New York: Charles Schribner's Sons, 1972.

Hall, Courtney Robert. *History of American Industrial Science.* New York: Library Publishers, 1954.

Hartog, Philip Joseph. "Smith, Robert Angus," *Dictionary of National Biography* 1885–1900, vol. 53: 112–14.

Heilbroner, Robert. *The Economic Transformation of America: 1600 to the Present.* Fort Worth: Harcourt Brace College Publishers, 1999.

Hellman, Hal. *Great Feuds in Technology: Ten of the Liveliest Disputes Ever.* (Wiley Popular Science). New York: John Wiley and Sons, Inc., 1998.

Hidy, Ralph W., and Muriel E. Hidy. *Pioneering in Big Business, 1882–1911: History of Standard Oil Company (New Jersey).* New York: Harper & Brothers, 1955.

Hill, Libby. *The Chicago River: A Natural and Unnatural History.* Chicago: Lake Claremont Press, 2000.

Hofstadter, Richard. *Social Darwinism in American Thought.* Boston: Beacon Press, 1955.

Hughes, Thomas Parke. *American Genesis.* New York: Penguin Books, 1989.

Johnson, George. *The Purification of Public Water Supplies*, Department of the Interior, United States Geological Survey, Water-Supply Paper 315, Washington: Government Printing Office, 1913.

Jones, Francis Arthur. *The Life Story of Thomas Alva Edison.* New York: Grosset & Dunlap Publishers, 1931.

Jonnes, Jill. *Empires of Light: Edison, Tesla, Westinghouse, and the Race to Electrify the World.* New York: Random House, 2003.

Josephson, Matthew. *Edison.* New York: McGraw-Hill, 1959.

Knowles, Ruth Sheldon. *The Greatest Gamblers: The Epic of American Oil Exploration.* New York: McGraw-Hill, 1959.

Larson, Henrietta M., Evelyn H. Knowlton, and Charles S. Popple. *New Horizons, 1927–1950: History of Standard Oil Company (New Jersey).* New York: Harper & Row, 1971.

Linsley, Judith Walker, Ellen Walker Rienstra, and Jo Ann Stiles. *Giant Under the Hill: History of the Spindletop Oil Discovery at Beaumont, Texas, in 1901.* Austin: Texas State Historical Association, 2002.

Lloyd, Henry Demarest. *Wealth Against Commonwealth.* New York: Harper & Brothers, 1894.

McDonald, John. *A Ghost's Memoir, The Making of Alfred P. Sloan's "My Years with General Motors."* Cambridge / London: MIT Press, 2002.

Maier, Pauline, Merritt Roe Smith, Alexander Keyssar, & Daniel J. Kevles. *Inventing America, A History of the United States,* 2 vols. New York: W. W. Norton & Company, Inc, 2003.

Markowitz, Gerald, and David Rosner. *Deceit and Denial: The Deadly Politics of Industrial Pollution.* (California / Milbank series on health and the public, 6). Berkeley: University of California Press, 2002.

Mayewski, Paul Andrew, and Frank White. *The Ice Chronicles: The Quest to Understand Global Climate Change.* Hanover, NH: University Press of New England, 2002.

Merchant, Carolyn. *The Columbia Guide to American Environmental History.* New York: Columbia University Press, 2002.

Melosi, Martin V. *Effluent America: Cities, Industry, Energy, and the Environment.* Pittsburgh: University of Pittsburgh Press, 2001.

Melosi, Martin V. *Thomas A. Edison and the Modernization of America.* Glenview, Illinois: Scott, Foresman/Little, Brown Higher Education, 1990.

Midgley, Thomas, IV. *From the Periodic Table to Production: The Biography of Thomas Midgley, Jr., the Inventor of Ethyl Gasoline and Freon Refrigerants.* Corona, California: Stargazer Publishing Co., 2001.

Millar, David, Ian, John & Margaret. *Cambridge Dictionary of Scientists.* New York: Cambridge University Press, 1996.

Misa, Thomas. *A Nation of Steel: The Making of Modern America 1865–1925.* Baltimore: The John Hopkins University Press, 1995.

Moran, Richard. *Executioner's Current: Thomas Edison, George Westinghouse, and the Invention of the Electric Chair.* New York: Knopf, 2002.

Morgan, H. Wayne. *Industrial America: The Environment and Social Problems, 1865–1920.* Chicago: Rand McNally College Publishing Company, 1974.

Mushet, Robert, Forester. *The Bessemer-Mushet Process, or Manufacture of Cheap Steel.* Cheltenham: J.J. Banks ,1883.

Nasaw, David. *Andrew Carnegie.* Reprint Edition. New York: Penguin Books, 2007.

Nevins, Allan. *John D. Rockefeller.* 1st ed. New York: Charles Scribner's Sons, 1959.

Ndiaye, Pap A. *Nylon and Bombs: DuPont and the March of Modern America.* Translated [from French] by Elborg Forster. Baltimore: The Johns Hopkins University Press, 2007.

Nye, David. *Electrifying America.* Cambridge, Massachusetts: The MIT Press, 1995.

Oliver, John W. *History of American Technology.* New York: The Ronald Press Company, 1956.

Pasachoff, Naomi. *Alexander Graham Bell: Making Connections.* New York: Oxford University Press, 1996.

Pittsford, William A."The Dollars and Cents about Smoke," *Proceedings of the Smoke Prevention Association of America*, 17th Annual Convention, 1923, pp. 27–30.

Porter, Glenn. *The Rise of Big Business.* Arlington Heights, Ill.: H. Davidson, 1973.

Pursell, Carroll, ed. *Technology in America: A History of Individuals and Ideas.* Cambridge, Massachusetts: The MIT Press, 1981.

Pursell, Carroll W., ed. *The Machine in America: A Social History of Technology.* 2nd ed. Baltimore: The Johns Hopkins University Press, 2007.

Quarateillo, Arlene Rodda. *Rachel Carson: A Biography.* (Greenwood Biographies). Westport, Conn.: Greenwood Press, 2004.

Rothenberg, Marc, ed. *The History of Science in the United States.* New York: Garland Publishing, Inc., 2001.

Seitz, Frederick. *The Cosmic Inventor: Reginald Aubrey Fessenden (1866–1932).* (Transactions of the American Philosophical Society, 89 part 6). Philadelphia: American Philosophical Society, 1999.

Silliman Jr., Benjamin. *Report on the Rock Oil, or Petroleum, from Venango Co., Pennsylvania,* 16 April 1855. Reproduced in the *American Chemist*, 2(1871–72): 18–23.

Smith, Robert Angus, *Air and Rain.* London: Longmans, Green, and Company, 1872.

Stover, John. *American Railroads.* Chicago: University of Chicago Press 1961.

Strouse, Jean. *Morgan: American Financier.* New York: Random House, 1999.

Sward, Keith. *The Legend of Henry Ford.* New York: Rinehart & Company, Inc., 1948.

Tarbell, Ida. *The History of Standard Oil Company.* Briefer Version. Mineola, New York: Dover Publications, 2003.

Travis, Anthony S. *Dyes Made in America, 1915–1980: The Calco Chemical Company, American Cyanamid and the Raritan River.* Jerusalem: Sidney M. Edelstein Center for the History and Philosophy of Science, Technology and Medicine at the Hebrew University of Jerusalem in conjunction with the Hexagon Press, 2004.

Twain, Mark. *A Connecticut Yankee in King Arthur's Court.* New York and London: Harper & Brothers Publishers, 1889.

Wall, Bennett H., C. Gerald Carpenter, Genes S. Yeager, and Earl N. Harbert. *Growth in a Changing Environment: A History of Standard Oil Company (New Jersey) Exxon Corporation 1950–1975.* New York: McGraw-Hill, 1989.

Wasik, John F. *The Merchant of Power: Samuel Insull, Thomas Edison, and the Creation of the Modern Metropolis.* New York: Palgrave Macmillan, 2006.

Williamson, Harold F. and Arnold R. Daum. *The American Petroleum Industry.* Evanston: Northwestern University Press, 1959.

Wise, Tad. *Tesla.* Atlanta, Georgia: Turner Publishing, Inc., 1994.

INDEX

NOTE: Page references in *italics* refer to figures.

A

Acheson, Edward, 130
acid rain, 132–137
Adams, James, 65
Adams, Julius, 114
Adee, Fred, 116
Aerial Experiment Association, 61
Air and Rain (Smith), 133
Air and Waste Management Association, 137
air pollution, 131–139
    acid rain and smoke, 132–137
    electrostatic precipitator invention, *138,* 138–139
    overview, 131–132
Alcoa, 27, 91–92
Alcock, Thomas, 125
Alexander III (Csar of Russia), 5
Alexander the Great, 106
Almer McAfee, 46
alternating current, 68–71, *69*
Aluminum Company, 24–25
aluminum industry, 24–30
    electrolytic production of aluminum, *24,* 24–29, *28*
    price of aluminum (1855-1914), 29
    U.S. production of (1883-1919), 30
American Bell Telephone Company, 57–58
American Bessemer, 14
American Chemical Society (ACS), 7
American Cyanamid Company, 132
American Institute of Chemical Engineers (AIChE), 7
American Locomotive Company, 89
American Railway Union, 101
American Smelting and Refining Company, 139
American Society for Metals (ASM International), 7
American Society for Testing and Materials (ASTM), 7
American Society for the Prevention of Cruelty to Animals, 72
American Speaking-Telephone Company, 55
American Sugar, 2
American Sugar Refining Company, 101
American Telegraph Company, 61
American Telephone & Telegraph (AT&T), 58, 93
American Tobacco Company, 2, 92, 96
American Waterworks Association, 129
American Waterworks Association, 129
Amoco, 96
Amy, Joseph, 106
Anaconda Copper, 139
Andrews, Samuel, 95
Aristotle, 106
Aryan Theosophic Society, 80

Atwood, Luther, 42
automobile industry, 43, 86–89, 131–132

B

bacteriology, 105–110
Baekeland, Leo, 85
Bailey, Marcellus, 53, 55
Baldwin, William, 53, 55
Bank of Goldschmidt, 28
Barnett Shale, 47
Barraud, Francis, 78
Barrett, William, 79
BASF, 45
Batchelor, Charles, 65, 75
Beau de Rochas, Alphonse Eugene, 87
Belgium Ministry of Public Works, 126
Bell, Alexander Graham, 1, 4, 6, 50–58, 59–61, 75
Bellamy, Edward, 98–99
Bell Telephone Company, 57, 62, 91–92
Benz, Karl, 86–87
Bergstresser, Charles, 2
Berliner, Emil, 62, 77
Berliner Gram-O-Phone Company, 77–78
Berthollet, Claude Louis, 8, 125
Bessemer, Henry, 10, 12–14, 15
Bessemer Steel Company or Association, 14
Bismarck, Otto, 76
Bissell, George H., 34–35
Black Maria studio, 79
Blackwell, S. H., 13
Blavatsky, Helena Petrova, 80
Board of Trade of Indianapolis, 137
Booth, Samuel, 34
borax industry, 83–84
British Aluminum Company, 29
British Perforated Paper Company, 117
British Petroleum (BP), 92
Brown, Harold P., 71, 72, 74
Brown, John, 11–12
Brown, Thomas, 13
Brunner Mond, 25
Brush Electric Light Company, 71
Brymbo Iron and Steel Works, 17
Buick Motor Company, 89

Bunsen, Robert, 24, 29, 34, 35
Bunsen photometer, 34
Bureau of Mines, 47
Burton, William, 45–46

C

California Borax Company, 83
Cambria Iron Works, 11, 14
Campbell, William, 116
capital punishment, electrocution and, 71–75, 73
Carborundum Chemical, 130
Carnegie, Andrew, 17, 91, 92, 93–94, 96–97, 99
Carnegie Corporation, 97, 130
Carnegie Institute of Technology, 97
Caruso, Enrico, 79
Castner, Hamilton, 24–25
Castner-Kellner Alkali Company, 25
Chapleau, Joseph, 72–73
Chelsea Water Works Company, 119
chemical industry, 82–85, 82–86
    borax industry, 83–84
    electrolytic industry, 85
    plastics industry, 85–86
    salt industry, 82–83
    sulfur industry, 84–85
Chemical Manufacturers Association (CMA), 7
Chesbrough, Ellis S., 113
Chevron, 96
Chicago Gas, 2
Chicago Hydraulic Company, 113
Chicago (Illinois)
    air pollution and, 132, 136–137
    water purification and, 113–116, 127, 130
Chick, Harriette, 128
chlorination, 124–130
cholera, 105–108, 109
Chrysler, Walter, 89
Chrysler Corporation, 89
Citizens Smoke Abatement League, 136
City Savings Bank of New Haven, 35
Clare, Thomas, 11–12

INDEX: 153

Clark, George, 1
Cleveland (Ohio)
    air pollution and, 134–135, 137
    water purification and, 110–115, 127, 130
Cleveland Water Company, 111
coagulation, water purification and, 122–123, *123*
coal age, 30, 31
Coleman, William Tell, 83–84
Coleman Company, 84
Columbia Phonograph Company, 79
Connecticut Yankee in King Arthur's Court, A (Twain), 99
Consolidated Talking Machine Company, 78
Continental Edison Company, 70
Coolidge, William D., 67
Cooper, Hewitt, and Company, 16
Corbett, "Gentleman" Jim, 79
Cornell, Ezra, 55
Cort, Henry, 8
Cottrell, Frederick, 138–139
Courtney, Peter, 79
Crapper, Thomas, 116
Crescent Hill Water Filtration Plant, 124
Crookes, William, 79
Csolgosz, Leon, 74
Cugnot, Nicolas, 86
Cullom, Shelby, 100
Cumberland Oil Company, 49
Cummings, Alexander, 115
Curie, Marie, 79
Curie, Pierre, 79

D

Daimler, Göttlieb, 86, 87
Daimler Motor Company, 87
Darby, Abraham, 8
Darby, James H., 17
Davis, Arthur Vining, 27
Davis, Edwin, 73–74
Davy, Humphry, 125
Day, David T., 46
Dayton Engineering Laboratories Company (Delco), 89

Debs, Eugene, 101
Dececo Company, 116
Department of Treasury, 129
Deutsche Bromoconvention, Die, 85
Development and Funding Company, 130
Devon Energy, 48
Diesel, Rudolf, 87
direct current dynamos, 27
Dodd, Samuel, 95
Doherty, Henry L., 49
Dow, Charles Henry, 2
Dow, Herbert H., 85
Dow Chemical Company, 85, 128
Dow Jones & Company, 2
Dow Jones Industrial Average, 2
Downs, James Cloyd, 128
Dow Process Company, 85
Doyle, Arthur Conan, 79
Drainage and Water Supply Commission, 114
Drake, Edwin, 31, 35–37
Drexel, Anthony J., 92
Drexel, Morgan & Co., 92
Dreyfus, Jules, 28
Duke, James B., 92, 97
Dunn, I. L., 49
DuPont Chemical, 128, 130, 132
Durant, William C., 88
Durfee, William, 11
Durfee, Zoheth, 11
Duryea, Charles, 87
Duryea, Frank, 87
dysentery, 106, 108, 109, 110, 113, 118, 130

E

E. C. Knight Company, 101
Eastman, George, 85
Ebbw Vale Iron Company, 13
Eberth, Carl Joseph, 104, 107, 108
economy, contractions and expansions of, 2–3
Edgar Thomson Steel Works, 93
Edison, Thomas Alva
    on alternating current, 71, 72
    on capital punishment and electrocution, 74–75

electrical industry and, 61–62
light bulb and, 64–67
motion picture machine, 79, 93
patents of, 1
phonograph, 75–76, *77*
rivals of, 55–57
spirit communication machine, 79–80
telephone contributions by, 62
Edison and Swan United Electric Light Company Limited (Ediswan), 64
Edison Electric Company, 66, 67
Edward VII, 93
electrical industry, 61–81
    alternating current, 68–71, *69*
    capital punishment and electrocution, 71–75, *73*
    General Electric inception, *66,* 66–68
    gramophone, 76–79, *78*
    light bulb, 63–65, *64, 65*
    motion picture machine, 79
    overview, 61–62
    phonograph, 75–76, *76, 77*
    spirit communication machine, 79–80
    technological breakthroughs (1850s-1920s), 81–82
    telephone and, 50–62, *52, 54, 59, 60*
    Westinghouse inception, 68–71
electrolytic industry, 85
electrostatic precipitator, *138,* 138–139
Elijah McCoy Manufacturing Company, 43
Enslow, Linn, 128–129
environmental issues
    air, 131–139 (*see also* air pollution)
    overview, 103–105
    steel industry and fossil fuels, 19–25
    water, 105–131 (*see also* water purification)
Environmental Protection Agency (EPA), 130
Essay on the Principle of Population (Malthus), 98
Eureka Iron Works, 12
Evans, Oliver, 86
Eveleth, Jonathan G., 34–35
external combustion engine, 86–87
Exxon Oil, 92, 96

F

Faraday, Michael, 1, 52, 119
Federal Reserve System, 93
Fell, George, 71, 74
Flagler, Henry M., 95
Flexner, Simon, 108
Ford, Henry, 86, 88
Ford Foundation, 97
Ford Motor Company, 88, 91–92
Frame, Robert, 116
Francisco, Antonio, 44
Frankland, Edward, 119
Frasch, Herman, 84–85
Fred Adee Company, 116
Frick, Henry Clay, 94
Fuller, George W., 124, 127

G

Gaffky, George, 108
Garfield, James, 59–60
Gayetty, Joseph, 117
Gay-Lussac, Louis Joseph, 32
Geissler, Heinrich, 63
General Bakelite Company, 85
General Education Board, 96
General Electric Company (GE), 2, 28, *66,* 66–68, 91–92, 93
General Motors Corporation (GMC), 88–89
George, Henry, 98
George Peabody & Company, 92
Gerry, Elbridge, 72
Gesner, Abraham, 33–34
Gibb, John, 118–119
Gilchrist, Percy, 16–17
Gilded Age, defined, 141
Gladstone, William E., 76
GMC Research Corporation, 88
Goddard, Robert, 139
Gold and Stock Telegraph Company, 61
Goodyear, 85
Goodyear, Charles, 3
Gordon, Bessie, 79
"Gospel of Wealth," 99
gramophone, 76–79, *78*

Gramophone Company (Victor Talking Machine Company), 79
Grange, The, 100
Gray, Elisha, 53, 55
Guiteau, Charles, 59, 60
Gulf Oil, 44, 46

H

Hall, Charles Martin, 1, 4, *24,* 24–28, *28,* 29
Hall, George, 27
Hall, Julia Brainerd, 27
Hamill, Al, 49
Hamill, Curt, 49
Harmony Borax Works, 84
Harrington, John, 115
Hart, John, 73
Hartog, Philip Joseph, 133–134
Hazen, Allen, 121
Helmholtz, Hermann von, 51
Henry, James T., 116
Hering, Rudolph, 114, 127
Hering and Fuller, 127
Hermany, Charles A., 124
Héroult, Paul Louis, 28–29
high-pressure catalytic cracking (crude petroleum), 45–46
Hill, David B., 71–72
His Master's Voice (Barraud), 78
Holley, Alexander, 14
Hooker Electrochemical Company, 130, 132
horizontal drilling (petroleum industry), 47–48
horizontal integration, 92
Houdry, Eugene, 46
Houdry Process Corporation, 46
Houston, Alexander, 126
Hubbard, Gardiner Greene, 53, 55, 57, 61
Hubbard, Mabel, 50–51
Humble Oil, 44
Humboldt, Alexander von, 32
Humphreys, Robert, 45
Hunt, Alfred, 27
Hunt, H. L., 44
Huntsman, Benjamin, 8
Hutchinson, Miller, 80
Hyatt, Isaiah Smith, 123
Hyatt, John Wesley, 123
hydraulic fracturing ("fracking"), 47–48

I

Ilosvay, Lajos, 134
Imperial Chemical Industries (ICI), 25
industrial organization, 91–101
　　characteristics of major industries, 91–92
　　industrialists who dominated American industry, 97
　　leading industrial organizers, 92–97
　　political reaction, 100
　　public reaction, 98–99
　　regulation, 96, 101
International Association for the Prevention of Smoke, 137
International Committee on Nomenclature, 107
Interstate Commerce Commission (ICC), 100
Iron and Steel Institute, 13
"Iron Road," 5

J

J. L. Mott Iron Works, 116
J. P. Morgan & Co., 92
Jeffries, James, 79
Jersey City Water Supply Company, 126
Jewett, Frank, 25
jigging down spring pole method (petroleum), 37–40, *38*
Johnson, Benhu, 110
Johnson, Eldridge, 78
Johnson, George A., 127
Joiner, Columbus Marion "Dad," 44
Jones, Bill, 94
Jones, Edward D., 2
Joule, James, 1
Joule's Law, 69
Joy Morton and Company, 82

K

Kelley, Oliver, 100
Kellner, Karl, 25, 79
Kelly, John, 9
Kelly, William, 8–12, 14, 15
Kelly Pneumatic Process Company, 11
Kemmler, William, 73–74
kerosene, 34–35
Kettering, Charles, 88–89
kicking down spring pole method (petroleum), 37–40, *39*
Kier, Samuel, 33–34
Kiernan News Agency, 2
Kirkwood, James, 114
Koch, Robert, 104, 107–108, 119, 126
Kruesi, John, 65

L

Labarraque, Antoine, 125
Laissez-faire, 3
Langley, Samuel, 61
Langmuir, Irving, 68
Lathrop, Austin, 72
Latimer, Lewis H., 65
Laura Spelman Rockefeller Memorial Foundation, 96
Lawrence, Ernest O., 139
Leal, John, 126–127
Lefferts, Marshall, 61
Lenoir, Etienne, 86
light bulb, 63–65, *64, 65*
Little, Arthur D., 7
Lloyd, Henry Demarest, 98, 99
Lodge, Oliver, 79
Lomonosov, Mikhail, 32
London Board of Health, 107
London Metropolitan Water Board, 126
Looking Backward (Bellamy), 98–99
Lord Kelvin (William Thomson), 79
Lösch, Friedrich K., 108
Louisville Board of Water Works, 124
Louisville Water Company, 124
Lucas, Anthony F., 44

M

Mabery, Charles F., 134–135
Malthus, Thomas, 98
Martin, Pierre E., 15–16
Maryland Department of Health, 128
Massachusetts State Board of Health, 121
Mathieson Alkali Works, 128
Maxwell Motor Company, 89
Maybach, Wilhelm, 87
McCoy, Elijah, 43
McGowan, Cecil, 126
McMillan, Daniel, 71
Meacham, George, 1
Medico-Legal Society, 72
Mellon family, 27
Mendeleev, Dmitri, 32
Metropolitan Sanitary District of Greater Chicago, 114
Michelson, Albert, 61, 85
Midgley, Thomas, 89
Midland Chemical Company, 85
Miller, Charles, 1
Mills, Hiram, 121
Mitchell, George, 48
Mitchell Energy and Development Corporation, 48
Mobil Oil, 44, 96
Moncrief, W. A. "Monty," 44
Monge, Gaspard, 8
Morgan, J. P. (Jay Pierpont), 92, 97
Morley, Edward, 85
Morrell, Daniel, 11
Morse, Samuel, 3
Morton, Joy, 82–83
Morton Arboretum, 83
Morton Salt Company, 82–83
motion picture machine, 79
Munn vs. Illinois, 100
Mushet, David, 11
Mushet, Robert, 11, 12–13

N

Napoleon III, 12
National Bell Company, 57

National Lead, 2
Neff, Charles, 116
Neward Filtering Company, 123
New York and New Haven Railroad
        Company, 35
New York Stock Exchange, 2
Niagara Falls Power Company, 70, 74–75
Nicholas II (Csar of Russia), 5
Nichols, William Ripley, 135
North American Review, 94

O

Occidental Petroleum, 130
Oersted, Hans Christian, 24
Ogden vs. Gibbons, 100
Ohm's Law, 74
Olds, Ransom, 86, 88
Oldsmobile, 88
Olds Motor Vehicle Company, 88
Olin Mathieson, 130
Orton, William, 55
Otis, Elisha, 67
Otis Brothers and Company, 67
Otto, Nikolaus, 86, 87
Owen, William Barry, 78

P

Pacific Coast Borax Company, 83, 84
Pacific Mail Steamship, 2
Pacini, Filippo, 104, 107
Pasteur, Louis, 107
patents, generally, 1. *see also individual
        names of industries and
        inventions*
Peacock, James, 106
Pedro II, Dom, 54
Pennsylvania Railroad Company, 93
Pennsylvania Rock Oil Company, 34, 35
Peterson, Frederick, 72, 74
petroleum industry, 30–49
        discovery, following Titusvil, 44
        early uses of petroleum, 32–35, *33*
        methods for finding and drilling (in
                nineteenth century), 37–40,
                38, 39, 40
        notable dates, 48, 49–50
        overview, 30–31, *31*
        production and price, 45
        storage and refining, 40–43, *41, 42, 43*
        thermal and high-pressure catalytic
                cracking of crude
                petroleum, 45–46
        Titusville discovery of petroleum,
                35–37, *36*
        U.S. petroleum reserves, 46–48
Pew, Arthur E., 46
phonograph, 75–76, *76, 77*
piston engine, 87
Pittsburgh Reduction Company, 27
Pittsburgh Water Works, 111
plastics industry, 85–86
Pneumatic Steel Association, 14
political issues, industrial organization and,
        100
pollution. *see* air pollution; water purification
Porzio, Luc Antonio, 106
Progress and Poverty (George), 98
Pupin, Michael, 6, 58

Q

Queen's County Oil Works, 42

R

Radio Corporation of America (RCA), 78
refineries (petroleum industry), 40–43, *41,
        42, 43*
regulation, industrial organization and, 96,
        101
Reithmann, Christian, 87
Reo Motor Company, 88
*Report on the Rock Oil, or Petroleum, from
        Venango Co. Pennsylvania*
        (Silliman), 34
Report on the Subject of Water Works
        (Cleveland, Ohio), 111
Research Foundation, 139
Richardson, Sid, 44
Richmond and Company, 82
Riverside Portland Cement Company, 139
Rockefeller, John D., 92, 94–96, 97, 99

Rockefeller, William, 95
Rockefeller Family Associates, 97
Rockefeller Foundation, 96–97
Rockefeller Institute for Medical Research, 96, 97
Roessler and Hasslacher Chemical Company, 128
Rogers, F. M., 45
Roosevelt, Theodore, 94, 101
Root, Elihu, 94
Royal Dutch Shell, 92

S

Safe Drinking Water Act, 130
Sainte-Claire Deville, Henri, 24, 29
salt industry, 82–83
Sanders, Thomas, 55, 57
sand filtration, *118,* 118–123, *119*
sanitation. *see* water purification
Schaudinn, Fritz, 108
Scheele, Karl, 125
Schwab, Charles M., 94
Scott, Clarence, 117
Scott, Edward, 117
Scott, Thomas, 117
Scott, Thomas A., 93
Scott Paper Company, 117
Scovill, Philo, 110–111
Scowden, Theodore R., 111, 124
Sedgwick, William, 121
Semmelweis, Ignaz, 125–126
Seneca Oil Company of Connecticut, 35
sewer systems, 110–124, *117, 118, 123*
Sharkey, Tom, 79
Sherman, John, 101
Sherman Antitrust Act, 96, 101
Shiga, Kiyoshi, 108
Sidgwick, Henry, 79
Siemens, Friedrich, 15–16
Silliman, Benjamin Jr., 34–35
Simpson, James, 119
Sloan, Alfred P., 88–89
Sloan-Kettering Institute for Cancer Research, 89
Smith, Francis Marion "Borax," 83, 84

Smith, Robert Angus, 132–133
Smith, W. Anderson, 133–134
Smith, William A. (Uncle Billy), 36
smoke, air pollution and, 132–137
Snow, John, 106–107, 126
Social Darwinism, 3, 94, 141–142
Société Electrométallurgique Français, 28
Société Métallurgique Suisse (later Aluminum Suisse), 28
Society of Chemical Industry, 25
Socony (Standard Oil of New York), 46
Solvay Company (Belgium), 25
Solvay Process Company (Virginia), 128
Sousa, John Phillips, 79
Southwick, Alfred, 71–72
Spencer, Herbert, 94, 98
spiegeleisen process, 12
spirit communication machine, 79–80
Sprengel, Hermann, 63
Standard Oil Company, 45–46, 84, 95–96, 99
Standard Sanitary Manufacturing Company, 116
Standard Works, 95
Stanley, William, 70
Starr, John W., 63
steel industry, 7–24
    cost and environmental impact, 18–24
    development of large-scale industry, 8–15
    open-hearth furnace process, 15–17, *16*
    overview, 7–8
    production volume, by countries, 19, 20–23
    specialty steel alloys, 18
Steel Patents Company, 14
sulfur industry, 84–85
Sulphur Bank Quicksilver Mining Company, 83
Sun Oil, 44, 46
Suwanee Furnace (Iron Works), 9–11
Swan, Joseph, 63–64
Swan Electric Light Company, 64
Swift, Gustavus, 91–92

T

Tabor, Erwin, 72–73
Taft, William Howard, 101
technological transformation, 142. *see also* air pollution; industrial organization; water purification
    in Europe and Asia (1870-1914), 5–6
    generally, 1–4
    stages of, 142
    in U.S., 6–7 (*see also* aluminum industry; automobile industry; chemical industry; electrical industry; steel industry)
telephone, invention of, 50–62, *52, 54, 59, 60. see also* Bell, Alexander Graham
Tesla, Nikola, 6, 70
Texaco, 44
thermal cracking (crude petroleum), 45–46
Thom, Robert, 119
Thomas, Sidney Gilchrist, 16–17
Thomson-Houston Company, 67
Thorp, Thomas, 133–134
Tiernan, Martin, 128
Time Warner, 92
Titusville, petroleum discovery in, 35–37
Titusville (Pennsylvania), petroleum discovery in, *36*
toilets, invention of, 115–117, *117*
Townsend, James, 35
transportation
    automobile industry, 43, 86–89, 131–132
    early petroleum industry, 41, *42*
    railroads, 5, 35, 93, 101
Trans-Siberian Railroad, 5
Tredegar Iron Works, 8
Trustees of the Pneumatic or Bessemer Process of Making Iron and Steel, 14
Twain, Mark, 94, 98, 99
Twyford, Thomas, 116
typhoid fever, 109, 113, 118, 120, 121, 124, 126, 130

U

Union Carbide and Carbon Company, 85
Union Forge, 9
Union Stock Yards, 127
United States Borax and Chemical Company, 84
United States Geological Survey, 46
United States vs. E. C. Knight Company, 100
United States Weather Bureau, 137
Upton, Francis, 65
U.S. Patent Office, 1
US Rubber, 2
US Steel Corporation, 93, 94, 130

V

Vacuum Oil Company, 46
Vail, Theodore, 58
Vandermonde, Alexandre, 8
Veatch, John A., 83
vertical integration, 91–92
Victoria (Queen of England), 132
Victor Talking Machine Company, 78
Vocal Physiology and Mechanics of Speech, 50–51
Volta Laboratory, 61

W

Wallace, Alfred Russell, 79
Wallace, Charles, 128
Wall Street Journal, 2
Ward, Eber Brock, 11
Waring, George, Jr., 114–115, 116
Warning, A (Edison), 71
water purification, 105–131
    chlorination, 124–130
    deadly waterborne diseases, 109
    industrial water pollution and, 130–131
    overview and bacteriology, 105–110
    water distribution and sewer systems, 110–124, *117, 118, 123*
Watson, Herbert E., 128
Watson, Thomas A., 51, 53–54, 57
Wealth Against Commonwealth (Lloyd), 99
"Wealth" (Carnegie), 94

Western Union Telegraph Company, 2, 55–56, 58, 61, 62, 75
Westinghouse, George, 68, 72, 73
Westinghouse Electric Company, 68–71, 70
Whitehead, Harry, 107
Wilhelm, Karl, 15
Williams, Charles, Jr., 51
Willys-Overland Company, 89
Windom, William, 100
Wöhler, Friedrich, 24, 25
Wolman, Abel, 128–129

Woodhead, Sims, 126
wood period (energy), 30–31
World Health Organization (WHO), 130

Y

Young, James, 33–34

Z

Ziegler, Matilda (Tillie), 73